D0231420

034011

COATINGS FOR CORROSION PREVENTION

OTHER BOOKS IN THE ASM

MATERIALS/METALWORKING TECHNOLOGY SERIES

Influence of Metallurgy on Machinability (1975)
Sulfide Inclusions in Steel (1975)
Influence of Metallurgy on Hole Making Operations (1977)
Control of Distortion and Residual Stress in Weldments (1977)
Nondestructive Evaluation in the Nuclear Industry (1978)
What Does the Charpy Test Really Tell Us? (1978)
Welding of HSLA (Microalloyed) Structural Steels (1978)
Prevention of Structural Failures (1978)

COATINGS FOR CORROSION PREVENTION

Papers presented at a symposium in the
1978 ASM Materials & Processing Congress

Philadelphia, Pennsylvania
November 9, 1978

Programmed by the
Cleaning, Finishing and Coating Division
American Society for Metals

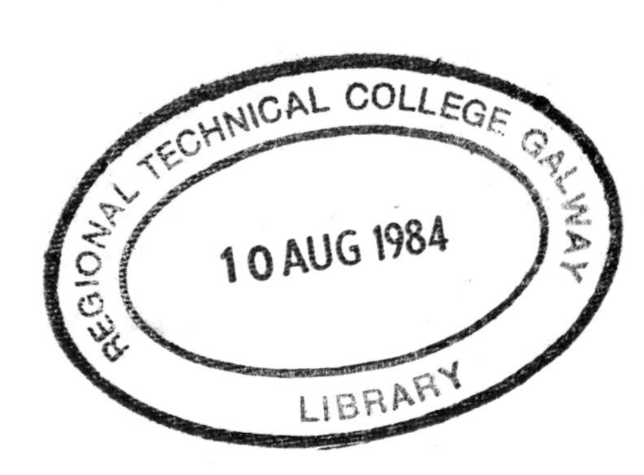

Materials/Metalworking Technology Series
AMERICAN SOCIETY FOR METALS
Metals Park, Ohio 44073

Nothing contained in this book is to be construed as a grant of any right of manufacture, sale, or use in connection with any method, process, apparatus, product, or composition, whether or not covered by letters patent or registered trademark, nor as a defense against liability for the infringement of letters patent or registered trademark.

Library of Congress Cataloging in Publication Data
Main entry under title:

Coatings for corrosion prevention.

(Materials/metalworking technology series)
Includes bibliographical references.
1. Protective coatings—Congresses. 2. Corrosion and anti-corrosives—Congresses. I. American Society for Metals. Cleaning, Finishing, and Coating Division.
II. Series.
TA418.76.C63 667'.9 79-17292
ISBN 0-87170-083-2

PRINTED IN THE UNITED STATES OF AMERICA

PREFACE

The 1978 ASM Materials & Processing Congress was programmed in its entirety by the Technical Divisions of the Society. In the case of the symposium addressed by these proceedings, the Coatings for Corrosion Prevention program was organized by the Cleaning, Finishing and Coating Division, William G. Wood, Kolene Corporation, Chairman.

Subject matter for the seminar was limited to papers relating to zinc and aluminum substrates and their top coatings. It is expected that subsequent efforts by the Division will address a wider range of organic and inorganic coating systems.

The generous cooperation of the authors and their companies is herewith acknowledged. Whatever compliments have been reflected on the Division for this effort rightfully belong to those who accepted and then carried out their assignment.

We wish to pay a particular acknowledgment to Dr. Frank LaQue for his outstanding contribution to the symposium both in his prepared remarks and in his contribution to the discussions.

William Cochran
Aluminum Company of America
Co-Chairman

Daryl E. Tonini
American Hot Dip Galvanizers Association
Co-Chairman

C O N T E N T S

CORROSION REACTIONS AS RELATED TO METALLIC COATINGS

F. L. LaQue

In view of my background, it seemed that I would be expected to deal with some aspects of corrosion in this opening session of the Cleaning, Finishing and Coating Division of ASM. In response to this, I am presenting a model which could be of use in understanding the behavior of coatings and, to some extent, in predicting the consequences of changes in metals used for coatings related to their electrochemical behavior, thickness, combinations, and so on.

The model is based on the premise that what happens at bare spots, pores or other discontinuities in coatings is likely to be more important than what happens to an intact coating thick enough to survive for a reasonable length of time in the use environment.

The model is based on using the several possible locations of the cathode in a corrosion reaction as shown in Figure 1.

As indicated, the cathode could exist at

1. The base of a single of multiple layer coating.

2. At the top of a multilayer system, e.g., chromium over nickel.

3. At the base of the top layer in a multilayer coating, e.g., chromium over nickel plating.

4. At the base of the second layer in a multilayer coating, e.g., under the bright nickel layer in a duplex nickel system.

5. Along the walls of a pore or bare spot in a coating.

6. On the surface of another metal in electrical contact with the coated metal in the same electrolyte, e.g., a dissimilar metal fastening.

It is generally agreed that corrosion is an electrochemical process associated with flow of current between surfaces having a difference in electrochemical potential. The surface having the higher potential acts as the anode in the corrosion cell while the surface having the lower, more noble, potential acts as the cathode.

The coatings can be less noble than the metal that is coated, e.g., zinc or aluminum on steel, and thereby make the steel the cathode in the corrosion reaction and protect it from corrosion by the electrochemical reaction.

Or coatings can be more noble than the metal that is coated or than some other metal in a multilayer system such as chromium over a duplex nickel system on steel, i.e., bright nickel over semi-bright, or levelling nickel, or nickel over copper.

What happens at discontinuities in cathodic coatings in accelerating attack of steel at the base of bare spots or other discontinuities can be the cause of more distress than the gradual loss of protection during the finite life of an anodic coating such as zinc or aluminum on steel. Consequently, the model that has been described will be used principally to help account for what happens with cathodic coatings such as chromium nickel plating on steel.

The current responsible for corrosion in a coating system cell conforms basically to Ohm's Law as shown in Equation 1.

However, the potential E driving the corrosion cell decreases with the flow of current. The potential of the anode drifts towards that of the cathode as a result of what is called anodic polarization Ap while the potential of the cathode drifts towards that of the anode by cathodic polarization, Cp. The extent of polarization increases with current density. As shown in Figure 2, polarization limits the amount of current and resulting corrosion that can occur with a particular combination of metals.

This limiting corrosion current is reduced further by the resistance of the corrosion cell circuit. The current that can flow in the presence of some value of resistance is determined by the IR drop that can be accommodated between the polarized potentials as shown in Figure 2.

As noted previously, polarization is a function of current density. If a cathodic area is reduced, the current density and the related cathodic polarization will be increased and the anodic corrosion current will be reduced. This will account for the better performance of micro-cracked or micro-porous chromium in which the cathodic area of chromium that can become involved with any particular area of underlying metal is much smaller than would be the case with a more limited number of anodic areas associated with the relatively large cathodic area of a substantially continuous surface of crack free or pore free chromium.

The working potential of a corrosion cell taking into account the effects of polarization is shown in Equation 2.

The overall electrical resistance of the corrosion cell circuit could include, (Equation 3)

1. The resistance of material occupying a pore or other discontinuity in a coating, this will increase with the thickness of the coating.

2. The resistance of the liquid or moisture film outside the pore.

3. The resistance of the coating metal, not likely to be significant.

Combining the factors determining the working potential and the components of resistance yields an overall equation (4) covering the corrosion current in a cell involving a metallic coating.

Let us now use this equation along with the possible locations of cathodes in Figure 1 to account for what happens in different situations.

If an important cathode exists at the base of a pore or other discontinuity in a coating the corrosion reaction will be limited by the fact that no large cathode can be associated with a relatively small anode in such a confined space. Therefore, the extent of cathodic polarization will be greater than if a large cathode were to be involved.

As shown in Equation (4) an increase in the factor Cp will reduce the corrosion current I.

This applies in a more limited way to a cathode at location (5) on the walls of a coating. The modest increase in the cathodic area of a thicker coating is not likely to have an important effect.

In the case of a pore extending to the base of the second layer in a three layer coating, e.g., bright nickel over levelling nickel or nickel over copper, it is very beneficial to have the outer layer less noble than the underlying layer. In this event, the cathodic underlying layer will be protected by the upper layer and the progress of a pit thru the upper layer will be arrested as shown by Figure 3.

This desirable result is achieved in duplex nickel systems when the presence of sulfur in the bright nickel layer makes it substantially less noble than the underlying layer of levelling nickel of lower sulfur content.

The opposite situation exists when the underlying layer is copper, more noble than, and therefore cathodic to both the upper layer of nickel and to the base metal, e.g., steel or zinc. A pit that starts in the upper layer of nickel can extend thru the copper layer and progress into an anodic base metal with undesirable consequences as shown in Figure (4).

The important galvanic relations among the different layers of a multiple layer plating system appear to be at the root of the advantage of the Cass and Corrodkote tests for chromium nickel plating as compared with the previously used salt spray tests.

The Cass and Corrodkote tests evidently disclose properly the favorable galvanic relationship between the layers in a double layer nickel plating system. This is demonstrated by the same patterns of attack in these tests, Figures 5 and 6, as occurs under natural service conditions, Figure 3.

When, as is usually the case, the principal cathodic reaction occurs on the general surface of a coating, the relatively large area of this surface as compared with that of the metal exposed at the bases of pores or other discontinuities will reduce the extent of the cathodic polarization factor in Equations 3 and 4 and thereby increase the danger of harmful corrosion currents.

Increasing the number of pores and thereby distributing the galvanic current so as to decrease the anodic current density would be theoretically advantageous from the standpoint of rate of penetration of the base metal. However, it would hardly be tolerable with respect to the decorative features of a coating with so many spots of rust or other corrosion products.

A more useful step would be to increase the thickness of the coating and to transfer the location of the critical cathode from the metal at the base of a coating to a more favorable location (4) between two layers of a coating, as in the case of duplex nickel systems.

The principal advantage of increasing the thickness of a cathodic coating would be to increase the electrical resistance of the material within a pore so as to reduce the corrosion current as per Equation 4. There would also be an advantage of reducing the number of pores thru a thicker coating so as to decrease the number of spots of disfiguring corrosion products.

The foregoing discussion relates only to coatings that are more noble than, and cathodic to, the base metal. In the case of less noble anodic coatings such as zinc or aluminum on steel the action is to insure that metal exposed at the base of pores will be made completely cathodic and thereby protected from corrosion by eliminating all anodic reactions. The operating potential in equation 4 becomes a desirable minus rather than positive factor.

The situation with a cathodic reaction on another metal surface is very much the same as with a cathode or anode on the surface of the coating. The principal factor will be the area of the dissimilar metal relative to that of the coating or the metal exposed at the base of the pores. The worst situation would be a relatively large area of another metal having a considerable potential cathodic to that of the base metal or the coating metal, or both.

In the special case of an anodized coating on aluminum the principal favorable effect will be to increase the electrical resistance of the corrosion cell. There will be an undesirable effect in situations involving a cathodic reaction on the surface of another metal. Here, the effect of the high electrical resistance of the anodized layer will be to concentrate and accelerate the more localized galvanic corrosion at the base of pores in the anodized coating. The same situation exists with organic coatings on the anodic member of a galvanic couple.

The electrical resistance of the electrolyte or film of moisture in contact with a coated metal can be an important factor.

The lower electrical resistance of sea water or salt spray decreases the factor (RL) in Equation 4. This will account for the greater severity of corrosive effects in marine environments as compared with fresh water or those well away from the sea and the greater likelihood of severe corrosion under conditions of immersion as compared with atmospheric exposure. It will account, also, for the more aggressive nature of polluted industrial environments as compared with the clean air in rural areas.

It is hoped that the model that has been presented and implications that may be drawn from it will be of some use in accounting for the behavior of coatings and leading to appropriate action to improve their performance.

EQUATIONS

EQUATION 1:

$$I = \frac{E}{R}$$

Where I = corrosion current; E= Difference in potential between anodic and cathodic surfaces; R = resistance of circuit

EQUATION 2:

$$I = \frac{(PA - Ap) - (PC + Cp)}{R}$$

Where I = corrosion current; PA = potential of anode; Ap = polarization of anode; PC = potential of cathode; Cp = polarization of cathode; R = resistance of circuit

EQUATION 3:

$$R = RDt + RL + RCt$$

Where R = resistance of circuit; t = thickness of coating; RD = resistance of material inside bare spot; RL = resistance of liquid outside bare spot; RC = resistance of coating material

EQUATION 4:

$$I = \frac{(PA - Ap) - (PC + Cp)}{RDt + RL + RCt}$$

Where I = corrosion current; PA = potential of anode; Ap = polarization of anode; PC = potential of cathode; Cp = polarization of cathode; RDt = resistance inside bare spot; RL = resistance of liquid outside bare spot; RCt = resistance of coating; t = thickness of coating

POSSIBLE LOCATIONS OF CATHODES
IN A METALLIC COATING SYSTEM

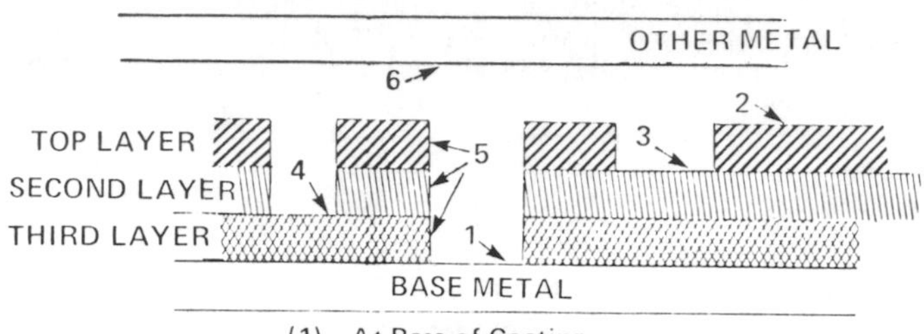

(1) At Base of Coating
(2) On Top of Coating
(3) At Base of First Layer
(4) At Base of Second Layer
(5) On Wall of Coating
(6) On An Other Metal Surface

FIGURE 1

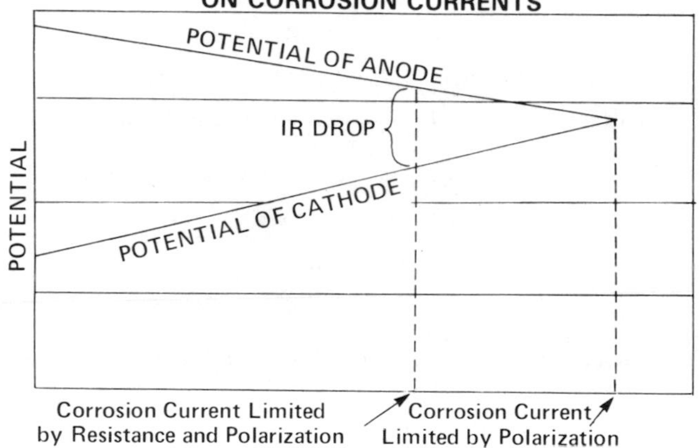

FIGURE 2

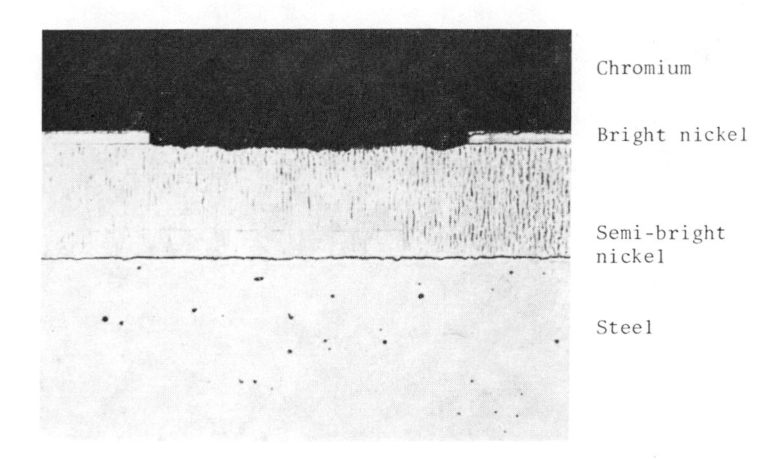

FIGURE 3. Pitting in double-layer nickel+chromium plated bumper bar after 29 months' service in an industrial environment. (X100)

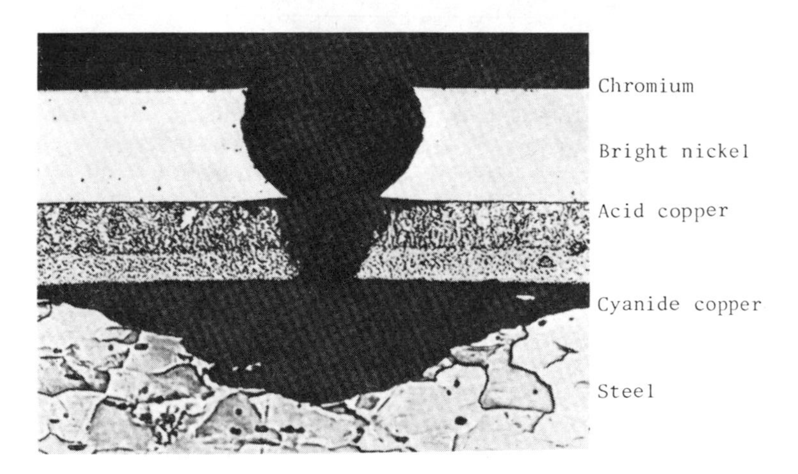

FIGURE 4. Pitting in copper+nickel+chromium plated bumper bar in region near exhaust pipe after 36 months' service in an industrial area. (X125)

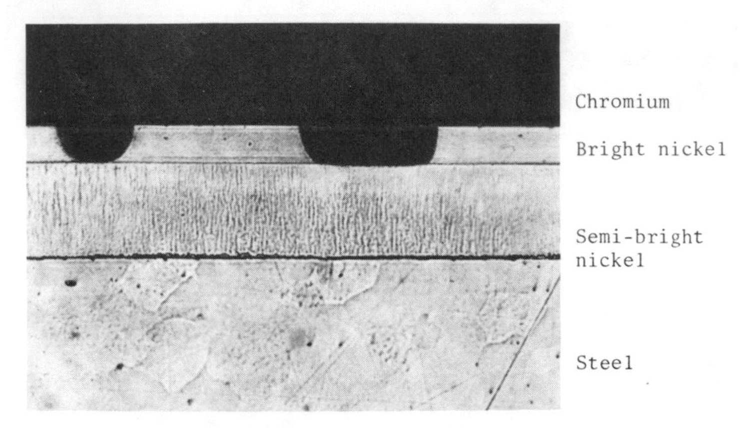

FIGURE 5. Pitting in double-layer nickel+chromium plated steel after 216 h exposure to acetic acid/ salt spray. (X175)

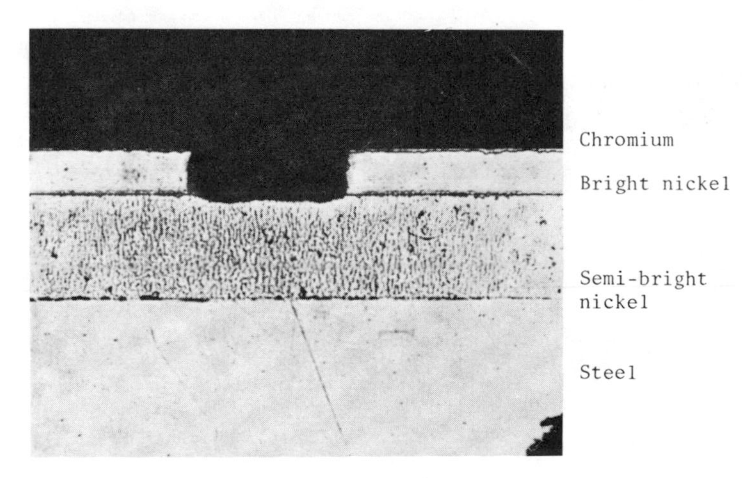

FIGURE 6. Pitting in double-layer nickel+chromium plated steel after exposure to 5 cycles of *Corrodkote* test. (X175)

ZINC-RICH PRECOATED STEEL

Griff W. Froman
Inland Steel Company

With the cost from the corrosion of cars and trucks in the United States alone estimated at $5.5 billion annually, corrosion prevention is probably the main task of materials engineers today. The most widely used methods for corrosion protection of steel involve coating the substrate with some sort of protective film, to either isolate the steel surface from the corrosive environment, or to sacrificially protect the steel by the preferential corrosion of the coating system.

Low-alloy plain carbon steels are commonly coated with aluminum, chromium, nickel or similar highly corrosion resistant metals to prevent the oxidation of the steel substrate. Such barrier films are highly effective corrosion preventive coatings unless the continuity of the film is disrupted during forming or by abrasion. In this case, the more noble coating acts as a large cathode of an electrolytic cell and in an aggressive environment actually accelerates the corrosion of the anodic steel base.

Galvanized coatings afford protection of the steel substrate by sacrificial means. Such films are anodic with respect to the steel and corrode in preference to the substrate. The degree of corrosion protection is directly related to the zinc film thickness and is only effective, of course, while metallic zinc is available.

By far the most commonly used protective coatings are based on the barrier film properties of paint systems. The total value of paints and lacquers produced in the United States amounts to about $2.5 billion annually, half of which is estimated to be used for corrosion prevention. All present-day paints are permeable in some degree to water and oxygen. Some resin systems are less permeable than others and, in general, their better performance as a diffusion barrier applies only to adherent multiple-coat applications that effectively seal up pores and other defects.

Zinc-rich paints represent an attempt to combine the sacrificial corrosion protection of zinc with the barrier film protection of paints. Finely divided metallic zinc particles are embedded in a resin matrix and even though the particles are covered by a thin film of resin a relatively good conductive path is produced by the point-to-point contact of the zinc dust. The electrical conductivity of this system allows the zinc to behave anodically to the steel substrate when a corrosive environment eventually permeates the paint film. Zinc pigmentation can be incorporated into either organic or inorganic binder systems. The inorganic films are usually quite brittle and are therefore applied as spray-on coatings to pre-assembled structures, whereas the organic zinc-rich paints exhibit the adhesion and flexibility required for prepainted applications.

Currently, U.S. coil coating facilities are capable of supplying approximately three million tons of precoated steel coils annually and several new coating lines are now under construction. Modern coil coating lines provide uniformity in coating film control and versatility in processing capabilities unique to the coatings industry. The schematic of a coil coating line, shown below, demonstrates the typical coil processing sequences available. Typically, steel coils are subjected to a variety of cleaning, brushing, and pretreatment stages prior to the application of the paint films. The coating rolls are normally operated turning opposite to strip travel with coating thickness controlled by roll speed, and film uniformity maintained by contact pressure. Curing oven capacity has been increasing so as to now accommodate line speeds over 500 feet per minute for steel strip close to 70 inches wide.

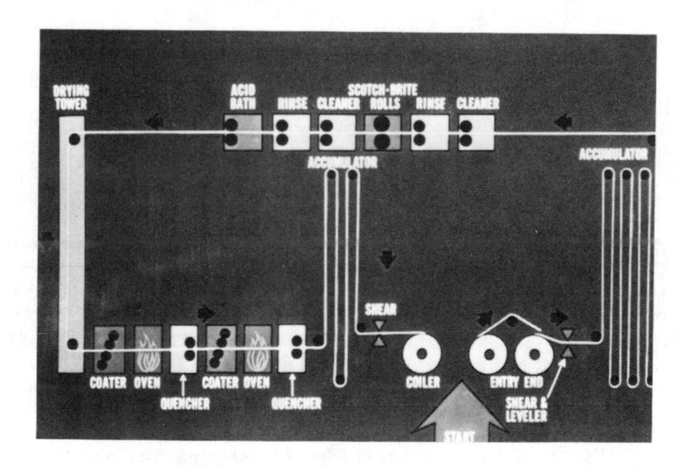

Fig. 1. Schematic of Coil Coating Line

This type of coil line has the capabilities for many prepainting operations which assures strip cleanliness and provides a variety of possible metal surface preparations. The coating rolls can be operated to paint both sides or only one side of the steel strip and can even be designed to paint only selected areas with no adverse effects on the uncoated steel surface characteristics.

A precoated product for use in a large market such as automotive must meet several important property requirements. Adequate corrosion resistance is, of course, very important but the coating must also have the adhesion and flexibility to withstand deformation when formed into sub-assembled parts by brake-press forming, roll forming, or stamping operations. Since the end user must further assemble these coated parts, the coating system must be weldable. The coating must also be compatible with conventional topcoat or exterior finish coatings. Zinc-rich paints have the capacity to meet these important material requirements.

A typical zinc-rich paint is produced by mixing very fine metallic zinc powder into a resin system to develop a complex film, as shown in the accompanying photomicrographs. The resin appears to cover and adhere to the zinc particles and thus binds them to the substrate as an integral film. A zinc dust size distribution is evident with a maximum particle size typically around 8 to 12 μm. The cross section micrograph shows the particle distribution to be also quite random throughout the thickness of the coating. The smaller particles tend to improve the packing efficiency of the dust thus increasing the number of contact points between particles and the overall contact area with the substrate. The packing efficiency is a function of particle size and concentration, which in turn affects the paint film conductivity.

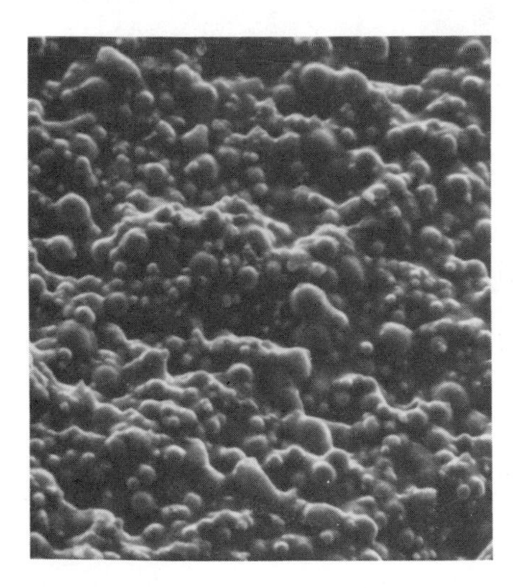

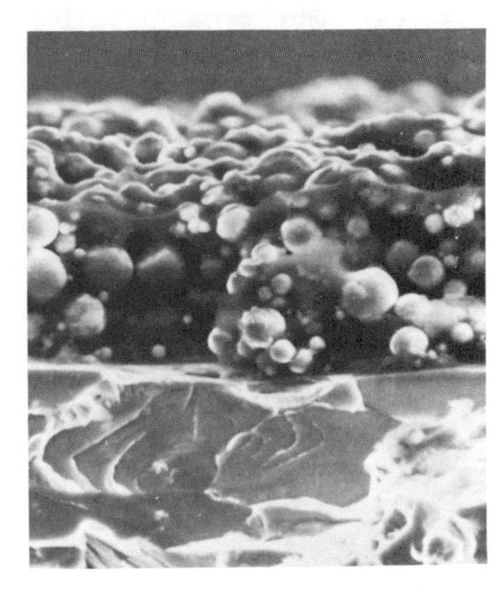

Fig. 2. Surface and Cross Section Photomicrographs of Zinc-Rich Paints

The most important factor contributing to the desirable performance properties of zinc-rich paint systems is the ability of the film to conduct electrical current. The resistivity of zinc-rich films as a function of zinc dust loading is shown below. Most zinc-rich systems operate at zinc loadings of between 80% and 95% by weight. This degree of metallic zinc pigment loading produces adequate particle contact which provides the electrically conductive paths for desired film conductivity.

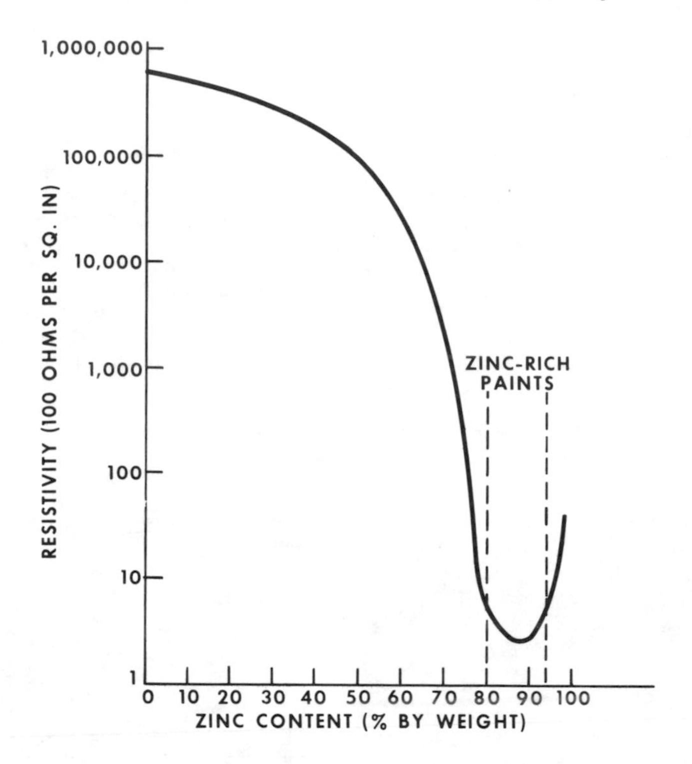

Fig. 3. Resistivity of Zinc-Rich Paints
as a Function of Zinc Content

With such electrically conductive paths present throughout the zinc-rich matrix this painted product has been shown to be weldable, a property which expands the application possibilities over conventional painted coil products. The attached weldability lobes of a typical zinc-rich coated steel, a plot of weld cycles versus applied current, show a rather large range of resistance spot welding conditions available to achieve an acceptable weld button.

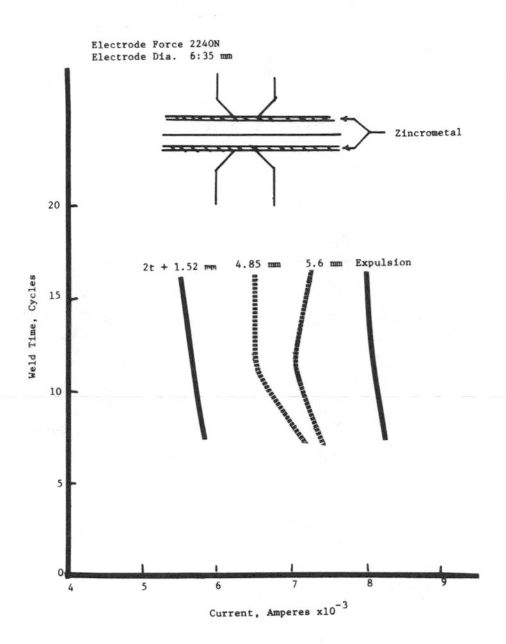

Fig. 4. Weldability Lobe
for Zinc-Rich Coated Steel

The weldability of a zinc-rich precoated system is also a function of the pretreatments or metal surface preparations required for any particular application. If a relatively non-conductive pretreatment such as a phosphate conversion coating is applied prior to the application of a zinc-rich paint, the resistivity of the entire precoated product is increased. Some pretreatment systems have been developed in which zinc, carbon, or other such electrically conductive pigments have been incorporated, thus maintaining an electrical coupling between the paint film and the substrate. The weldability lobes for the different surface preparations reflect the system's ability to conduct electricity, i.e., broader welding parameters are possible for the new conductive films, and a rather limited welding range is typical of the more insulative coating systems.

The electrical conductivity of the zinc-rich system also has an effect on the corrosion protective mechanism provided by the coating. The highly conductive systems exhibit a greater degree of white zinc corrosion products on the surface of test panels exposed to 5% salt spray exposure than do systems with lower conductivity. The prevention of red rust corrosion from the iron substrate at scribed or severely formed areas is also more evident with the more conductive systems. The more integral electrical coupling of the metallic zinc dust in the paint to the steel substrate allows for a greater degree of anodic or sacrificial protection. The less conductive systems provide a lesser degree of sacrificial protection at exposed areas but yet still provide general barrier film protection.

Anodic polarization measurements, shown for some typical materials in Figure 5, illustrate the different corrosion mechanisms. Conductive zinc-rich systems composed of paints applied directly to the steel substrate, or in conjunction with a conductive pretreatment, exhibit similar passivation behavior when immersed in 0.5 N sodium sulfate solution. The same zinc-rich paint applied over a phosphate conversion coating exhibits a behavior expected of an insulating layer with limited electrolyte permeability. The open circuit potentials of the conductive systems are very similar to those of hot-dip galvanized coatings, and the amount of current flow at 100 mv over open circuit may indicate the extent of local corrosion cells set up within the system by the contact between the dissimilar metals. The lower current values at open circuit for the more electrically resistant system demonstrates the lack of these sacrificial corrosion reactions. It is hoped that further work with such polarization techniques would lend more insight to the corrosion mechanisms of zinc-rich systems, and aid in determining long-term corrosion protection.

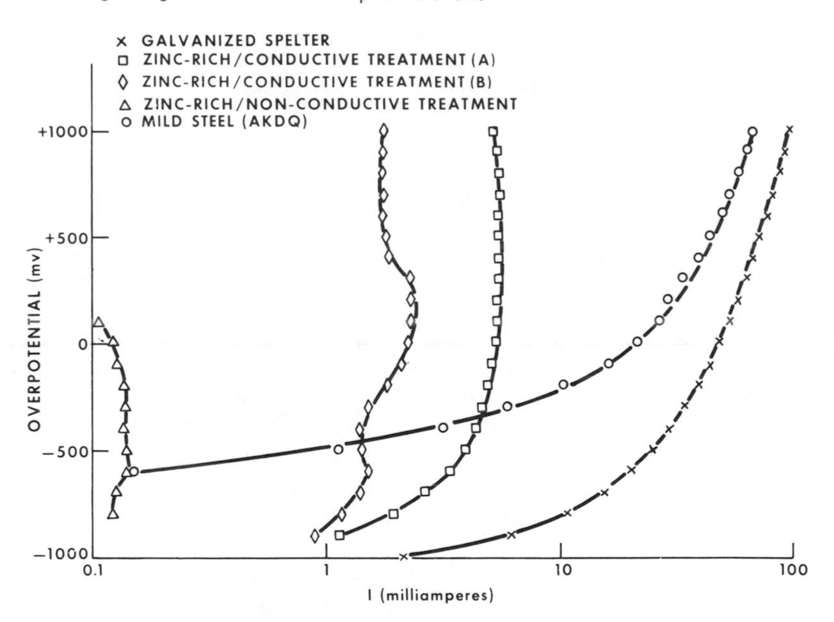

Fig. 5. Anodic Polarization Curves of Cold-Rolled, Galvanized and Zinc-Rich Coated Steels

The comparison of zinc-rich paints to hot-dip galvanized products is unavoidable since both systems incorporate metallic zinc to protect an iron substrate. Such comparisons however can be very misleading because the products are substantially different in many respects. Zinc-rich paints usually contain only about one quarter as much zinc per unit area as does 1-1/4 oz. (G.90) hot-dip galvanized steel. The paint offers barrier film properties, incorporates inhibitive pigments which passivate the zinc dust and steel substrate, and zinc-rich paints are not designed to withstand exposure to ultraviolet light sources. A comparison of salt spray corrosion resistance however does give some indication of the corrosion protection provided by zinc-rich systems, as shown in Table 1.

Table 1. Comparison of Salt Spray Corrosion Resistance in Terms of Percentage of Red Rust Coverage

Material	12 hours	240 hours	500 hours
Zinc-rich paints	0	0	<1
HDG, 1-1/4 oz.	0	25-50	50+
Zn/Fe alloy	0	90+	100
Cold-rolled steel	100	100	100

The adhesion of zinc-rich systems is good when tested by routine bend, impact, or Olsen button deformation, or even when subjected to stamping die operations. No loss in film adhesion is noted even in severely drawn areas limited by the formability of the base steel. Because the coating does not flake or pick off in the stamping dies, scoring or scratching problems are minimized. As a result of the application uniformity obtainable by coil coating of zinc-rich paints, print-through defects do not occur on critical exposed surfaces. Shown in Figure 6 is a typical example of a complex automotive part demonstrating the film integrity even on such severely deformed structures. Similar stamped sections exhibited no red rusting after 240 hours in 5% salt spray tests at the deep drawn corner areas.

Fig. 6. An Example of a Sub-Assembly Illustrating the Formability and Weldability of Zinc-Rich Precoated Steel

The formed zinc-rich coated parts are resistant to hot alkali spray cleaners and rinsing cycles incorporated into most post-painting processes. Zinc-rich paints are compatible with a variety of finish coat paint systems, providing good adhesion and improved underfilm corrosion resistance. The conductivity of zinc-rich primers allows the deposition of both anodic and cathodic electrodeposited paints now in use throughout the finishing industry. Laboratory evaluations with such duplex coatings show improved performance after salt spray exposure compared to electropaints applied to cleaned and phosphated steel, as shown in Figure 7. The advantage of this duplex system is in providing assurance of corrosion protection to even severely recessed areas of parts if such areas limit adequate electrophoretic deposition.

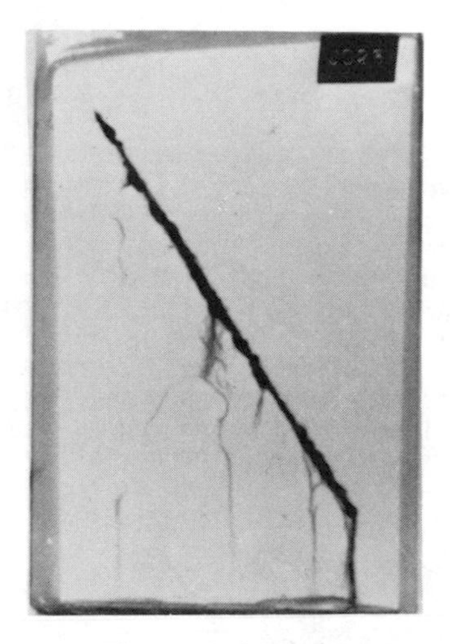

Electropaint over
cold-rolled steel

Electropaint over
zinc-rich paint

Fig. 7. Salt Spray Corrosion Performance
of Electrodeposited Paints

The major producer of a zinc-rich system is Diamond Shamrock Corporation who developed Zincrometal®. Zincrometal® is a two-coat system comprised of a Dacromet® base coat and a Zincromet® top coat. Dacromet® is a proprietary combination of chromic acid, zinc dust and other chemicals which, when cured, provides the conductivity and passivation for the Zincromet® top coat zinc-rich primer. The growth of Zincrometal® precoated steel over the past few years has been remarkable. From its first commercial applications in 1969, Zincrometal® production has reached an annual level of over a million tons.

By far the largest market for zinc-rich precoated steel has been the automotive industry with their concerns of vehicle corrosion from increased road salt usage. Because of the outstanding properties of the zinc-rich systems and the versatility of the coil coating application process, other market areas are being investigated. Different applications may not require the same coating properties deemed necessary for the automotive industry. Zinc-rich systems can be tailored to produce the desired properties for other specific end uses. Automotive manufacturers require a highly critical one-side product with optimum property performance. Other major industrial applications may not require such stringent welding parameters and would desire a two-side coated material as a corrosion resistant primer for exposed decorative finish paints. Such an application could then use a phosphate pretreatment applied in the wet section of the coil line with a one-coat zinc-rich system which would still be weldable and yet provide an excellent corrosion resistant paint base.

Improved zinc-rich paints are required to maintain the production growth of such systems. The coating property requirements, of course, must be met or improved by any new system. But a single-coat zinc-rich paint applied directly to the steel substrate would provide a great advantage by eliminating the present double-cure cycle and thereby reduce gas consumption at the coil line. Low cure temperature zinc-rich systems are a major goal not only for energy considerations and increased coater line speeds, but the reduced peak metal temperature may be required for the new high-strength steels which may be more temperature sensitive than the aluminum-killed drawing quality steels commonly used today. Paint systems may have to incorporate a catalyzed cure mechanism to achieve such low temperature requirements and the versatility of the coil coating lines make this approach feasible. Even the use of different metallic pigments or combinations of pigments should be considered to achieve the basic performance properties.

Zinc-rich paints have been proven to be a corrosion resistant coating system that can be welded, formed, cleaned and topcoated with the versatility and control which makes zinc-rich precoated steel unique to the coatings industry.

REFERENCES

(1) H. H. Uhlig, Corrosion and Corrosion Control, 2d ed., John Wiley & Sons, Inc., New York, 1971.

(2) R. Burns and W. Bradley, Protective Coatings for Metals, Reinhold, New York, 1967.

(3) K. B. Tator, "Zinc-Rich Coatings Forge to Leadership in Corrosion Control," Proceedings of the National Zinc-Rich Coatings Conference,

(3) conducted by the Zinc Institute, Inc. (December 4, 1974).

(4) A. W. Kennedy, "You Can Fool Mother Nature with Zincrometal," Proceedings of the National Zinc-Rich Coatings Conference, conducted by the Zinc Institute, Inc. (December 4, 1974).

(5) C. A. Dahlke, Jr., "Zinc-Rich Primers: Choice and Selection," Materials Protection and Performance, X, No. 9 (September, 1971).

ZINC COATINGS FOR CHEMICAL PLANTS & REFINERIES
THREE STUDIES IN DECISION MAKING

Glen Nishimura
Zinc Institute, Incorporated

INTRODUCTION

At the 1977 NACE Canadian Region Western Conference, Terry Lawry[1] stated that "In specific regard to corrosion protection by coatings in the construction industry, we, the 'experts', are not achieving what has to be achieved."

Several faults in which corrosion engineers were deemed to be remiss were listed in that paper. These faults can best be summarized as a lack of technology transfer. In its simplest form, technology transfer means that you allow someone to benefit from your experience.

This paper is an attempt at technology transfer from firms now using zinc coatings for the protection of structural steel to firms anticipating decisions on protective coating systems. And, in fact, as will be shown in the paper, the decision of these firms to use zinc coatings was largely based on the transfer of technology from other firms or from parent organizations.

THE DECISION MAKERS

Three manufacturers of chemical and petroleum products were interviewed by Zinc Institute personnel. Each of these firms had recently completed or are presently engaged in major capital construction projects in which substantial quantities of zinc coated steel are used. The purpose of the interviews was to document the decision making process. Texaco, Inc. and Canadian Fertilizers Industries were two. The third asked to remain unidentified.

Texaco, Inc. has installed approximately 2700 tons of galvanized steel at its new 1430 acre refinery at Nanticoke, Ontario. The function of the galvanized coating is to protect Texaco's initial investment, to keep maintenance costs to a minimum and at the same time, to prevent any downtime in the production units due to corrosion of structural steel. A breakdown of the total galvanized weight shows that about 1870 tons can be found in columns, beams, "I" & "L" frames, while 130 tons were used for platform structures, stairways, and for some 700 ladders that connect different levels of the refinery. The remaining 700 tons were used for platform steel, handrails and other structural portions.

Six hundred and eighty tons of fabricated columns, beams, purlins, girts and fasteners were galvanized for use in the Canadian Fertilizers Industries urea/fertilizer complex in Medicine Hat, Alberta. The main storage warehouse and four auxiliary buildings-pumphouse, utility building, compressor building and maintenance building were all constructed with the above building components.

In addition the third Canadian producer of chemicals and refined products interviewed has recently erected 1780 tons of structurals and fabrications for new plant construction and modifications to existing facilities.

Each of the interviews dealt with three questions:

1. Where do you use galvanized steel?

2. How does the cost to galvanize compare with other coating systems?

3. Do you paint galvanized steel?

These questions were asked in several ways to represent different points of view. No attempt was made to channel the responses to fit a designed questionnaire, and no attempt was made to develop detailed data on the use or performance of galvanized steel. However, these simple, straightforward questions and the spontaneous answers yielded a remarkably clear picture of the use of galvanizing in that portion of chemical and petroleum industries represented by the three manufacturers.

INTERVIEW RESULTS

It was found that selection of a protective system involved a qualitative judgement on the relative importance of many factors that can vary widely according to the type of structure, its function, its general location, its immediate environment, and any changes (natural or otherwise) that may occur in its environment.

Each of the companies interviewed showed extensive use of galvanized steel. Nine specific general applications for galvanizing were cited: structural steel, pipe supports, pipe, grating, platforms, ladders, vessels, railings, and fasteners.

One company indicated that they apply a paint topcoat to galvanized steel in chemically aggressive areas. In all other instances, the three firms rely on the ability of the galvanized coating to protect the structure. For maintenance purposes each of the companies spot prime damaged or depleted areas and two companies paint the entire structure to prevent further deterioration.

All of the companies buy to predetermined specifications and all inspect the material that is received.

Welding galvanized steel posed a minor problem for each of the companies and bolted connections using galvanized fasteners was a preferred practice.

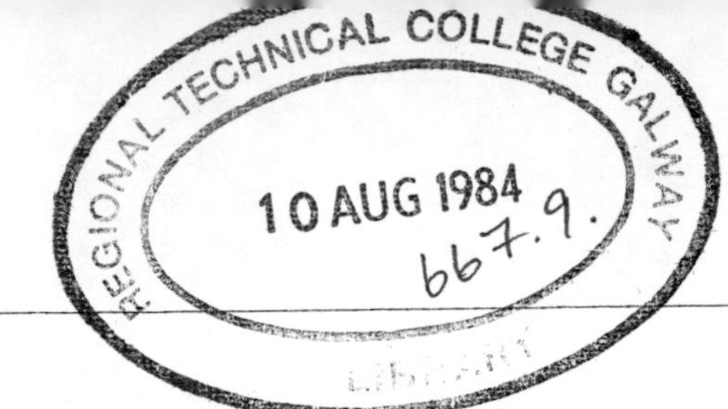

Two companies cited a formula based on the ratio of surface area to weight for the steel items as a determining factor in their choice between galvanized steel and zinc-rich paint. This formula, pioneered by a commercial galvanizer in the USA and described by Vickers[2] is used when steel structures are to be zinc coated and there is an option between applying a zinc-rich paint or hot dip galvanizing. It assumes equal corrosion resistance for both coatings.

CRITERIA FOR CHOICE OF COATINGS

In today's complex world, the corrosion engineer might expect a profound and weighty answer to the question "Why do you specify galvanized steel?" However, in each instance, the answer was simple and concise. It was not an answer that was based on volumes of sophisticated laboratory data nor on extensive field testing of comparative coatings.

The specification of hot dip galvanizing for coating structural steel is an empirical decision which is based on the collective experience of each company and its personnel with the coating. The companies reported that they preferred zinc coatings to any other coating and therefore use it whenever it is possible to obtain it because of long life, low maintenance, and low cost per year of life. The type of zinc coating used is influenced by several factors. If the article is complex in shape, if it has surfaces that are not fully accessible for blast cleaning and spraying, hot dip galvanizing is preferable to either zinc-rich coating or metallizing. If the article is larger than a conveniently located galvanizing facility can accommodate, one of the sprayed coating systems is the obvious solution.

An important factor in the selection of galvanized steel was a formula based on the area-to-weight ratio. As stated above, this formula is based on the assumption that galvanized steel and zinc-rich paint are capable of offering the same protection for equal coating thickness. All the companies interviewed use zinc in one form or another. Therefore, it appears that among coatings poeple who know the benefits that accrue from zinc coatings, formulae based on the ratio of surface area to weight are an important consideration.

A 20 YEAR CASE HISTORY

What I have shown is that one firm decides to use zinc because of the experience of another firm or a parent organization. To quantify this experience on which some of these decisions are based, let me use a fourth example--that of a 20 year old polyethylene plant owned by Soltex Polymer Corporation in Deer Park, Texas. The Gulf Coast of the United States is a very formidable environment because the normal corrosiveness of an industrial atmosphere is combined with the effects of high ambient temperatures, high relative humidity, some salt air, and severe tropical storms.

In 1967 T.F. Shaffer[3] reported on the condition of the galvanized steel after 10 years of service in a Celanese Corp. plant situated on the Gulf Coast. In January 1978, ZI personnel conducted a 10 year follow-up to that paper to determine the condition of the zinc coating after 20 years of service.

The plant at Deer Park, Texas is now owned and operated by Soltex Polymer Corporation, a subsidiary of Solvay & Cie of Belgium.

Galvanizing has been a standard coating for the Deer Park operation since the plant was constructed in 1956. That portion of the plant, which is reported in Shaffer's 1967 paper, was temporarily shut down in mid 1975 because it was energy intensive. However, the plant environment is essentially unchanged. This case history will discuss the three eras of construction which constitute the plant today.

1956 - 1978

Celanese Plastics Company based its decision to use galvanized structural steel in its polyethylene plant at Deer Park on the expectations of low cost per year of life and high performance. Empirical data from another plant belonging to the same company in Texas had shown that galvanizing performed very well with no maintenance. A ten to fifteen year life was estimated for the proposed hot dip galvanized coating.

The uses of galvanized steel in the plant include wide flange beams, standard I-beams, structural angles, channels, floor plates, pipes, stairways, ladders, railings, and miscellaneous items. With few exceptions, all the bolts and hardware on these structures are also hot dip galvanized.

In October 1966, it was reported that "Without exception, the galvanized structural steel is in excellent condition. A few rust spots were observed on the threads of anchor bolts attaching the base plate of vertical structurals to concrete footings. Occasional traces of alloy stain can be found. However, except for the anchor bolt threads, there was no rust visible on any galvanized structural steel in the plant."

Coating thickness measurements in 1966 averaged 5 to 7 mils with a high of 15 mils on some structural members and a low of 3 mils on 2½ x 2½ x ¼ angle.

In January 1978, thickness measurements on structural columns, stairway stringers, and beams averaged 5 mils after 20 years. The zinc coating on the top of a railing which saw very heavy traffic measured 0.7 mils on the sides and 2 mils on the center. The galvanized coating on fasteners was gone and pitting of the steel had occured. The structures, however, were not in distress and the galvanized coating adjacent to the fasteners was not depleted.

1962 - 1975

All structurals in the plant built during this period were hot dip galvanized except for one area which was field painted. Galvanizing was used for all platform steel, pipe racks, and structural supports. In addition, the rail car loading facilities and all interior steel in process buildings was galvanized. An inspection of that portion of the plant which had been field painted with a two coat system revealed that the coating was failing after 5 years.

1978

A polypropylene plant is currently under construction. Here again, all structural steel both indoors and outdoors is hot dip galvanized. Color identification of process streams and critical areas is achieved by painted stencils on the galvanized fabrication instead of painting the entire item.

The galvanized structural steel in the Soltex plant has lived up to its owners highest expectations. Galvanizing has eliminated the need for concern about maintenance painting and it has assured a neat and clean-looking plant even during periods when operating costs are under severe pressure.

CONCLUSION

The specification of zinc coatings for structural steel is an empirical decision which is based on the collective experience of companies and their personnel. Three Canadian companies reported that they preferred zinc coatings to any other coating because of low cost per year of life, low maintenance, and long life.

An inspection of an all galvanized polyethylene plant of the Solvay Corporation after 20 years of service has shown that hot dip galvanized steel is a highly favored material for plant construction. At the Soltex Corporation plant the galvanized steel has been maintenance free for 20 years, and in spite of the harsh industrial and marine atmosphere of the Gulf Coast, the zinc coating is expected to last many more years.

REFERENCES

(1) B.T. Lawry, "Some Observations and Opinions on Corrosion Protection by Coatings in the Construction Industry," NACE Canadian Region Western Conference, February 1977.

(2) R. Vickers, "Combination Coating System: Hot Dip Galvanizing and Inorganic Zincs can give Economical Protection," Materials Protection III, (8) August 1964.

(3) T.F. Shaffer, "The Use of Hot Dip Galvanized Steel in Chemical Plant Construction," 8th International Conference on Hot Dip Galvanizing, June 1967.

ONE-SIDE ELECTRO-GALVANIZING THE SIMPLE APPROACH

Charles L. Faust, Ph. D.
Thomas A. Sellitto
Allen W. White

In recent years, automobile manufacturers and others have established a requirement for drawing quality steel having good corrosion resistance on one side and a surface on the other side suitable for high gloss paint application. The preferred material is coated on one side with a heavy zinc coating by a process that does not change the metallurgical property of the sheet.

Production Machinery Corporation believes that hot dipped galvanized steels, "Zincrometal" coated steels, non-ferrous metals and plastics all have their place in these surface panels; however, we believe that steels electro-galvanized on one side are becoming stronger candidates for many of these applications. This belief is based on an extensive study made of each of the other contending materials for these critical surface applications, and concluding that each has a characteristic that makes it undesirable in specific applications.

For the purpose of this paper we are presenting only galvanizing techniques, not "Zincrometal" or plastics.

HOT DIPPING METHODS

Existing hot dipping methods were carefully evaluated and found to have shortcomings for some applications requiring one side coatings.

The simplest approach required additional equipment to apply a stop-off coating to prevent zinc adherence to one side of the strip and later this could be removed. However, the stop-off coating could not be removed easily and totally for subsequent paint application, nor was it 100% effective in preventing the molten zinc from adhering.

Hot dip galvanizing must be done with strip temperatures above 800° F. which ages the metal and adversely effecting its drawing quality.

Another hot dipping method is to pass the strip over a standing wave of molten zinc. This scheme also has drawbacks. Because of the thin coating used in automobile body panels, it is difficult to obtain the close coating thickness control required. The surface tension of the zinc may also cause it to throw-around to the uncoated side. Additionally, the thermal cycle is almost identical to that experienced in conventional hot dip lines, and therfore drawing properties of the steel can be adversely effected.

Another approach for one side coating, using hot dipping techniques is to hot dip the material and electro-chemically or mechanically deplate one side. In addition to the problem of overheating the strip, it is a costly approach and jeopardizes the bare steel surface quality.

A fourth hot dip type approach, although not truly one side, is to use differential coatings from one side to the other of the strip. This can conceivably be used for some body panel material if the additional cost of the zinc could be justified. At the present time, we suggest that this is a more expensive process and may be limited to special applications similar to those relegated for conventional two side coatings.

ELECTRO-DEPOSITION METHODS

Existing electro-plating techniques were also evaluated, and found lacking in many respects. However, there were no theoretical shortcomings such as the aging process caused by high temperatures experienced in the hot dip lines and "Zincrometal" lines.

Horizontal plating cells, using slab zinc or zinc ball anodes, have been used to produce one-side coating with minimum throw-around. However, the following problems do exist.

A. Quality. Zinc coating was somewhat porous, reducing its sacrificial behavior below the desired minimums.

B. Production. The strip had to be floated on top of the electrolyte presenting a serious challenge to the production department.

C. Cost. The zinc dissolved into solution faster than it could be plated out, necessitating disposal of part of the electrolyte on a constant basis. Obviously, this increased the operating cost and created a need for heavy metal disposal.

D. Quality. Iron was dissolved by the electrolyte. This caused sludging problems and also adversely effected the quality of the plating if allowed to continue.

E. Quality. Foreign particles from the consumable anodes dropped onto the strip and passed between the conductor rolls causing undesirable sheet embossing.

F. Quality. The low current density (200-300 amps/square foot) necessited large installations and could also result in porous plating.

G. Cost. Because of the large spacing, approximately 2", between the anode and strip, power requirements were relatively high.

H. Production. The soluble zinc anodes required appreciable manpower just to maintain a constant strip to anode gap as well as to replace them as they were used.

Another electro-deposition method using large drum-type conductor rolls and wrapping the strip around these to seal one side away from the solution was evaluated.

This process does produce high quality one side coated product; however, for the following reasons it was found undesirable during this study. a.) Maintenance of the seals on these large conductor rolls could be very high, increasing the operating cost. b.) Tensions in the plating cell are extremely high to assure conformance of the strip to the roll, resulting in larger drives and increase operating costs. c.) The plating currents within each cell are so high that it requires special current collectors, again resulting in high maintenance and high operating costs. d.) Initial capital equipment costs are high. e.) Chloride solutions used are extremely corrosive, resulting in special machine design, high maintenance, and increase cost.

Even though there are many problems associated with electro-deposition techniques, we have determined it to be the most feasible approach for one side zinc coating because of the inherent metallurgical problems with the high temperature operation of a hot dip or "Zincrometal" line. We also concluded that the product must be superior in performance, yet cost competitive with all other coatings.

DESIGN AND DEVELOPMENT

To maximize the efficiency of an electro-plating line, P/M set the following objectives:

A. To produce a dense coated product for maximum corrosion resistance at minimum coating thickness.
B. To increase operating performance and reduce operating costs.
C. To simplify the line and minimize equipment costs.
D. To have a system useful for both new and existing plating lines.
E. To minimize chemical effluents.

Our first concern is high zinc density to provide maximum corrosion protection at minimum coating thickness. The most dense zinc plate is achieved at high current densities such as 600-1000 ASF.

The single most serious problem in achieving high densities in an electro-deposition cell is the polarization effect as shown in figure 1.

FIGURE I POLARIZATION

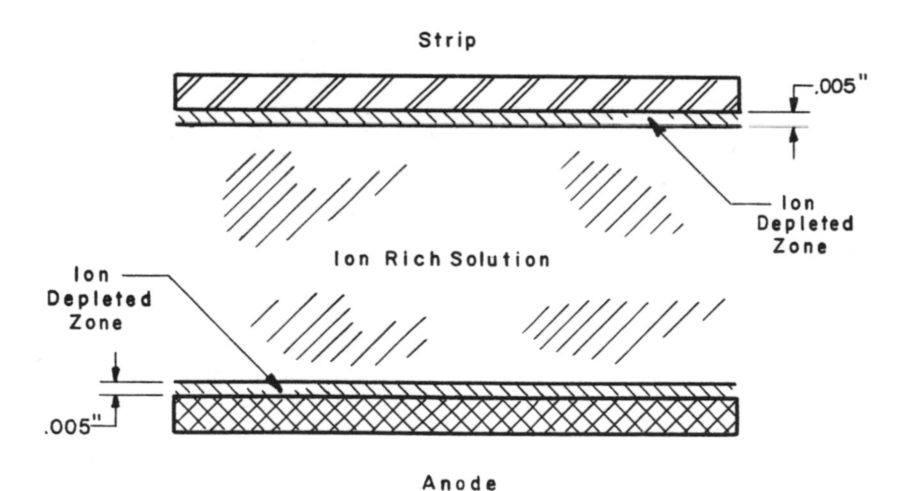

Production Machinery Corporation conceived a system to reduce this ion-depleted zone to less than 20% of its traditional thickness, by creating turbulent electrolyte flow at the strip, as shown in figure 2.

FIGURE 2 TURBULENT FLOW

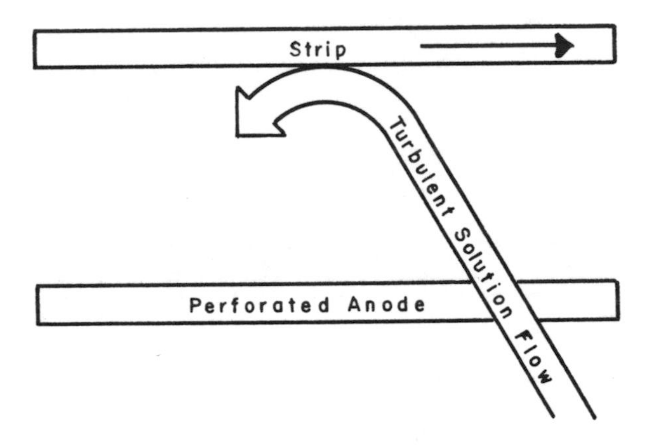

Another serious problem in achieving high densities is the generation of large quantities of gas at the anode and the cathode. Our new system effectively removes these gas bubbles, by using this same turbulent flow.

To increase operating performance and reduce operating costs, plating cell operating labor must be reduced or eliminated. The soluble anodes used on conventional plating lines require frequent adjustment and replacement. Insoluble anodes will virtually eliminate plating cell labor. Furthermore, they reduce solids contamination that causes surface damage to the product.

Material selection for these insoluble anodes was made on the basis of current carrying characteristics, resistance to damage by the solution, and costs. Several materials are available that work successfully.

The actual design of the anode is somewhat more complex than the material selection. The normal current path, as shown in figure 3, was analyzed as an electrical field.

FIGURE 3 NORMAL CURRENT PATH

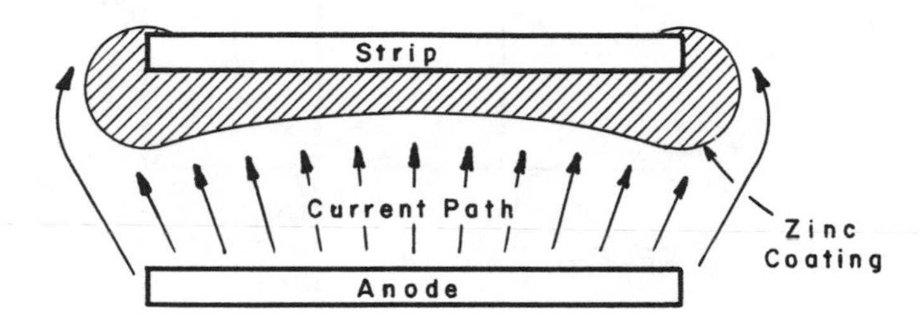

FIGURE 4 ANODE

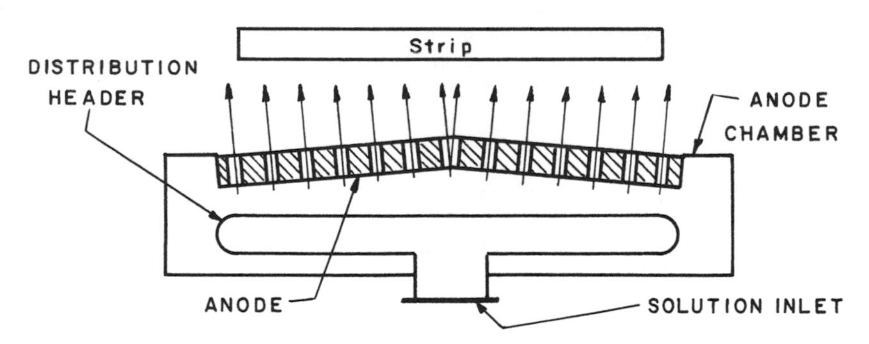

By curving the anode surface away from the strip from centerline to its edges it was found that more even current densities, and therefore coating thickness are achieved. Another benefit resulted from this curvature is that the strip to anode distance is greatest at the strip edge where the greatest clearances for wavy edges is desired. The anodes are shown in figure 4.

FIGURE 5

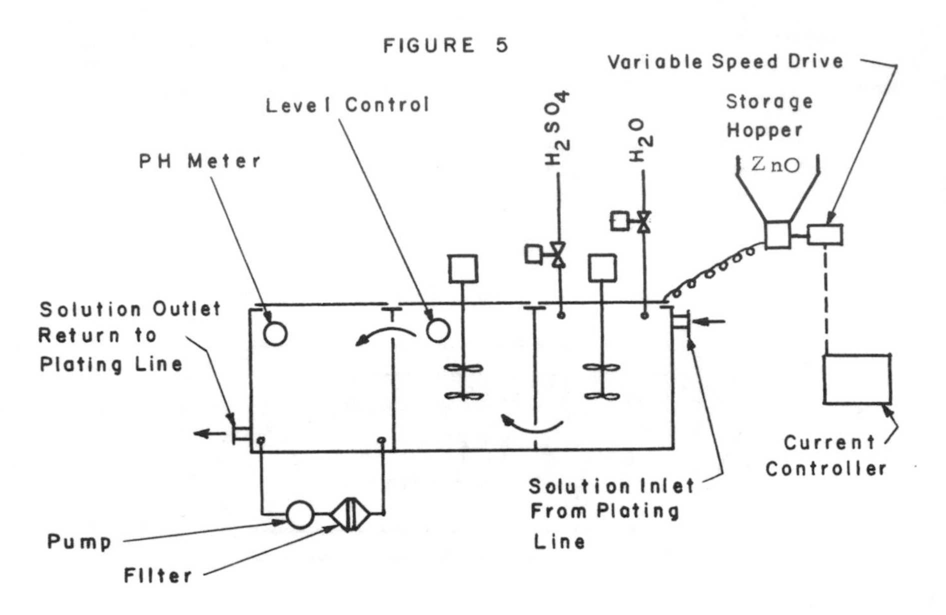

Since we determined that insoluble anodes must be used, a source of zinc had to be found. There are existing systems which utilize dross as a zinc source, or slab zinc can be electro-chemically dissolved. Either of these could be used with this plating process; however, dross may not be available and dissolving slab zinc is an energy user. Zinc is commercially available as an oxide, carbonate or hydroxide in addition to as a metal. Zinc oxide powder

The powder is added to the solution on a continuous basis at a solution preparation station located off-line, see figure 5. The feed rate is varied in direct proportion to the plating rate. The feed rate can be biased to compensate for any losses due to dragout.

Sulfuric acid and water are also added at the solution preparation station in very small quantities because of the losses due to dragout.

The third item of importance is to increase the operating performance and reduce operating costs. Two costs are major; plating power and personnel. The easiest way to reduce electrical power costs is to reduce plating voltage by minimizing the anode to strip spacing. Close spacing can be achieved by the use of the insoluble anodes. The first line will have 1" spacing. Early trials will be run to optimize this spacing. Figure 6 shows voltages in a theoretical cell using this 1" spacing. Personnel costs are reduced by use of insoluble anodes because the plating cells need only minor attention.

FIGURE 6 PLATING CHARACTERISTICS

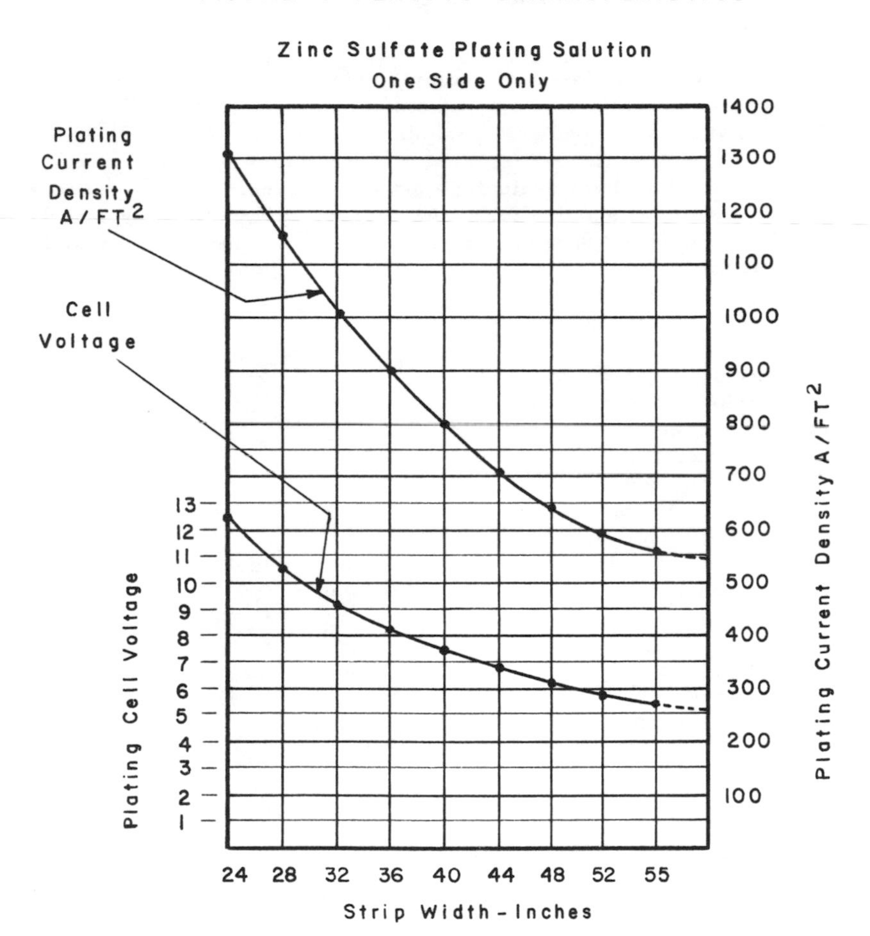

By using horizontal plating cells, the plating section can be used in many existing lines simply by removing the existing cells and installing the new ones. This can reduce the overall capital equipment expenditure and installation time.

Equipment costs were evaluated at each step during design and development. All materials were kept standard and manufacturing and assembly practices were evaluated. The basic plating section is floor mounted over a shallow pit. Rectifiers are self-contained, water cooled types, located immediately behind each cell to minimize bus bar requirements.

Using counter current rinsing techniques, the line meets all EPA regulations and is almost completely a closed loop system. The electrolytic cleaner is dumped on a periodic basis.

Operating costs have been evaluated on a typical line. The following operating costs are based on the following average parameters for a 72" wide line with a maximum speed of 350 FPM. Typical line arrangement is shown in figure 7.

AVERAGE PARAMETERS

Average Strip Width	44"
Average Strip Thickness	.030"
Average Line Speed	322 FPM
Average Product Yield	96%
Average Line Utilization	84%
Average Plating Efficiency	90%
Average Coating Thickness	.0005"
Average Coating Weight	.3 ounces/square foot - one side (G-60)
Average Current Density	720 amps/square foot
Average Voltage	10 Volts - DC
Total Current	500,000 amperes
Total Current/Tray	25,000 amperes
Number of Trays	20
Number of Contact Rolls	21
Average Strip Weight	1.25 lbs./square foot

The following is a summary of the costs, excluding depreciation and building costs.

Plating Electricity	$180.00/hr.
Cleaner Electricity	$ 10.00/hr.
Terminal Equipment Electricity	$ 60.00/hr.

Alkali	$ 11.50/hr.
Water	$ 1.50/hr.
Sulfuric Acid	Initial charge of sulfuric acid is $9,000.00. Losses due to drag-out are negligible and are not considered.
Steam Dryers	$ 5.50/hr.
Steam Cleaner	$ 5.50/hr.
Zinc Oxide	$600.00/hr.
Labor	$100.00/hr.
Total	$974.00/hr.

The above costs are based on commercially available services at rates determined in January, 1978. At these operating costs, the process will be competitive with the single side zincrometal and hot dip coatings.

SUMMARY

The first commercial facility is now being installed and should be in production by early 1979. It is a two cell line and will operate at a maximum speed of 350 FPM.

This line is simple to install, simplet to operate, low in cost and a non-polluter. It could be used by a manufacturing plant, toll coater, or service center as well as in a steel mill.

We feel that the product will meet all surface critieria and metallurgical requirements and because of the low cost will find wide acceptance, not only in the automotive industry but in other industries requiring corrosion protection on one side and paintability on the other side.

"Autogalv" is covered by several pending patents in the United States.

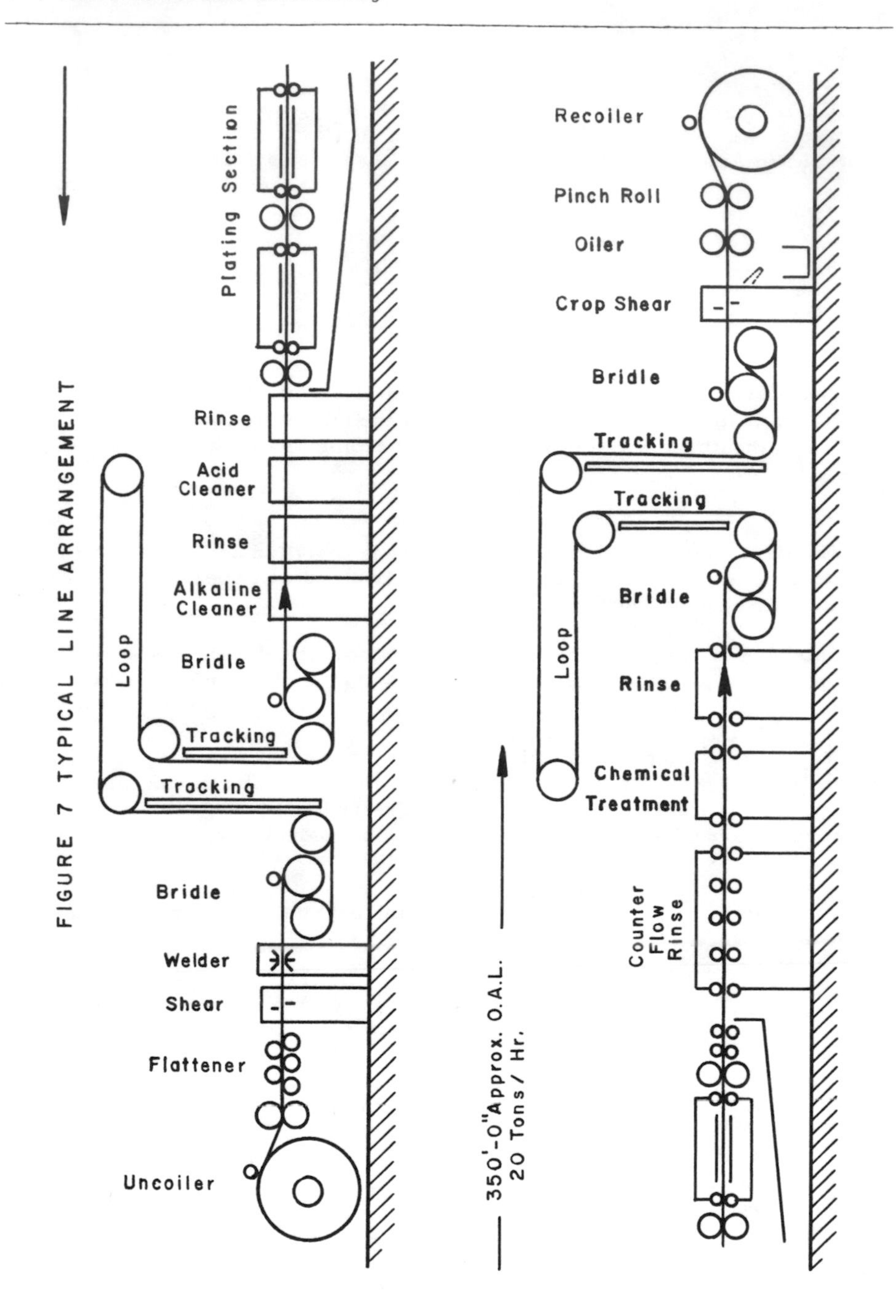

FIGURE 7 TYPICAL LINE ARRANGEMENT

MECHANICAL PLATING

E. A. Davis

Many industries are under tremendous pressures these days - with demands from customers pulling one way and demands from government agencies pulling the other. Many of these divergent demands may be in themselves legitimate, but their total effect creates that tremendous pressure to design better and faster at lower cost.

While there are many problems to solve, and no one solution is suitable in itself, improved corrosion resistance in small parts is one answer that helps satisfy two demands: customer satisfaction and cost reduction.

The process of mechanical plating has been used by various industries and their suppliers as a low-cost, efficient way to plate fasteners, stampings, extrusions, springs and many other small, exposed parts. These manufacturers have found mechanical plating highly satisfactory; not only does it eliminate many of the disadvantages of other finishes, but it has inherent advantages to contribute.

This paper will describe mechanical plating and point out those advantages, opening the way for manufacturers to apply them to their finish requirements.

The original system employed a series of dip tanks for surface preparation with the primary specialized portion of the process being the actual metal powder plating, done in a horizontal closed barrel. The system had limited process control which caused reliability inadequacies. Surface preparation, operator handling and equipment limitations dictated the necessity to provide easier-to-control technology.

Phase II brought about improved deposition on difficult-to-plate parts, improved surface preparation capabilities and refined handling techniques, characterized by single-use surface preparation chemicals to replace the dip tanks. However, one drawback was a more costly operation with increased energy consumption. Although reliability was extended beyond all previous capabilities, and plating techniques were expanded to include previously unplated substrates, cost and water consumption demanded additional refinements and improvements.

Continued research and development brought about more modern and current mechanical plating procedures. The present single-step process is easy to control and use, had no costly intermediate rinse cycles, and has simplified effluent treatment control to conform with local and national governmental regulations. All guesswork has been removed from the process; a specific amount of chemistry is used to plate a given amount of surface area. Detailed, easy-to-read charts and graphs provide the operator with the necessary data to quickly and simply plate many different part types to any predetermined thickness.

WHAT IS MECHANICAL PLATING?

Mechanical Plating is the cold welding of a ductile metal onto a metal substrate by the use of mechanical energy. The process is based on the fact that if an oxide free metal is placed in contact with another oxide free metal, metallic bonding can take place because of the free exchange of electrons.

HOW DOES MECHANICAL PLATING WORK?

There are six basic items necessary to perform the process:

1) Multi-sided, lined barrel.
2) Metal powder.
3) Glass bead impact media.
4) Oil free parts.
5) Proprietary chemistry.
6) Room temperature water.

In the Mechanical Plating process, finely divided metal powder is charged into a multi-sided, lined barrel with glass bead impact media, oil free parts, proprietary chemistry, and a small amount of water. The barrel is rotated at various speeds dependent on part size and configuration. Rotational energy generated by the turning barrel is transferred to the parts through the sliding/tumbling action of the glass bead impact media. The proprietary chemistry produces and maintains oxide free surfaces of the substrate and also of the metal powder to be plated. The impact media ensures good contact between the metal powder and parts to be plated and facilitates plating of inner recesses of various part configurations. It is the glass bead media that actually peens the metal powder onto the substrate to form the sacrificial coating.

WHAT METAL POWDERS CAN BE PLATED BY THE PROCESS?

Typically, metals that are Mechanically Plated are ductile - the most common are Cadmium, Tin, and Zinc as well as various combinations of these powders. Other metals that have been Mechanically Plated are Lead, Indium, Silver, Copper, Brass and Tin/Lead solder, However, the demand for plating the last group of metal powders has not been large.

WHAT METALS CAN BE PROTECTED BY MECHANICAL PLATING?

1) Low carbon steel.
2) High carbon heat-treated steel.
3) Cold rolled steel.
4) Various tool steels.
5) High strength alloys.
6) Some stainless steels.
7) Free Machining steels (including leaded steels).
8) Nitrided steels.
9) Sintered iron (powder metallurgy).
10) Some malleable iron and nodular iron.
11) Cast iron.
12) Copper.
13) Brass.
14) Bronze.
15) Lead.
16) Zinc Diecastings.

Mechanical plating is a process for parts that are normally and typically batch handled, having no greater dimension than 6 in. (152.4mm) in length and weighing less than 1/2 lb. (.227kg) each. Significant strides have been made toward plating and handling of larger and heavier part types; however, the majority of parts currently being Mechanically Plated are of the smaller variety.

PLATING PROCEDURES

Process Steps:

1) Parts are thoroughly degreased and rinsed.
2) Parts and impact media are loaded into plating barrel.
3) Water level and temperature are adjusted to allow a small pool of room temperature water to appear just ahead of the tumbling load.
4) Barrel rotational speed is set.
5) First surface conditioner for oxide, scale and rust removal is added.
6) Second surface conditioner is added for coppering step. *As previously discussed, many different types of substrates are coated. Various types of metal substrates will receive plating at different rates. By adding a copper strike, the metal powder does not see the substrate - only the copper - and the rate of deposition can be more closely controlled. Subsequent coating adhesion is also enhanced.*

7) Plating promoter is added. *This chemical is added to insure complete and even metal powder distribution to all surfaces within a given time cycle and to prevent lumpy or thin coatings (part to part variation).*
8) 'Flash' metal is added. *A small amount of metal powder (1 to 3 ounces per 100 sq. ft. or 28 to 85 grams per 9.3 sq. mtr. of surface area) is added to produce a thin coating completely covering each part. It has been determined that this 'flash' coating is beneficial in providing a tight, adherent bond between the substrate and the protective plating metal.*
9) Plating metal is added. *Plating metal is added in one or two increments depending on desired thickness and part configuration.*
10) Parts, media and spent plating chemicals are discharged. After discharge, the plater barrel is re-charged and the next load started. Parts and media are either screen or magnetically separated, and the media is rinsed and put in stand-by for the following run. The parts are post plated treated, dried, and packaged for shipment. Spent plating chemicals are pumped to a holding reservoir for subsequent waste treatment. Time sequence for a complete plating cycle runs between 40 and 55 minutes.

ADVANTAGES AND BENEFITS

Mechanical Plating Provides:

1) Assured product reliability through elimination of hydrogen embrittlement. *Studies of coating methods for hardened steel parts have shown that embrittlement caused by exposure to acid or other environments in which the steel is randomly anodic and cathodic is of a transient nature and can be relieved either by a 24-48 hour room temperature holding period or by traditional baking procedures. Finishing methods such as electroplating in which the steel parts are constantly cathodic, however, can produce irreparable damage to the parts. No amount of time nor baking can alleviate the damage. Mass spectrometry studies indicate that no correlation between evolvable hydrogen content and propensity toward brittle failure of hardened steel parts exists.*
2) Mechanical cleaning action from the media which definitely aids in scale removal and shortens the time of the preplate cycle.
3) Ability to plate tangling and flat parts completely and evenly. The impact media tends to keep tangling parts and flat parts which normally mask from making contact with others, allowing complete plating deposition.
4) A system that consumes all chemistry during each process cycle eliminating bath maintenance.
5) Ability to change metal coatings from batch to batch in the same equipment.

6) Non-complex waste treatment procedures; no cyanides or chelating agents are used in the process.
7) Ability to plate powder metallurgy without impregnation. A positive aspect is the ability to obtain good plating deposition on powder metallurgy without costly impregnation. Powder Metallurgy is an ever increasing industrial art in which metallic powders are manufactured, and articles made from them. The metal powders are pressed into formed objects which are subsequently heated to a high temperature to produce a coalesced, sintered, alloyed mass. In electroplating, the porosity of powder metallurgy parts is responsible for the entrapment of cleaning and plating solutions. The pores act as thermal pumps, forcing the solutions into the part interior or to the outside, depending on the temperature difference between plating bath and part. These solutions often break down, leaving potentially harmful deposits that eventually bleed out to ruin the finish and form premature corrosion products. Porous surfaces also make it difficult to obtain smooth or continuous electroplating. Therefore, it becomes necessary to impregnate the powder metallurgy parts with resins, plastics, or wax before they can be electroplated. The impregnation step is costly and time-consuming, but it is not necessary prior to mechanical plating. In fact, mechanical plating of powder metallurgy is basically the same as plating of other metals. Once it has been established that the powder metallurgy part is of sufficient density, is oil-free, and is not impregnated, normal procedures are followed.
8) Attractive economics especially for coatings in the 0.0003" (0.0076mm) and higher range.
9) Good adhesion and uniformity of coating on interior and exterior of part.
10) Ability to deposit heavy coatings in the 0.002" to 0.003" (0.05 to 0.076mm) range. This process is called Mechanical Galvanizing and has application in many industries where heavier coatings are specified. The difference between Mechanical Plating and Mechanical Galvanizing is basic:

    Mechanical Plating deposits a protective metal coating onto a substrate in a thickness range from 0.0001 to 0.0007 inch (0.00245 to 0.0177mm). Mechanical Galvanizing plates substrates in thickness ranges from 0.00075 to 0.003 inch (0.019 to 0.0762mm). The chemistry and procedure used for the Mechanical Galvanizing finish is similar to Mechanical Plating; however, plating cycle times are 15 to 30 minutes longer per run because metal powder deposits are up to 10 times heavier than normal coatings. The heavier deposits necessitate a longer cycle time to achieve complete plating consolidation. The advantages of Mechanical Galvanizing follow closely those of Mechanical Plating; the merits of the Mechanical Galvanizing process are as follows:

A) Batch-to-batch and part-to-part uniformity.
B) Simple and easy control of plating thickness.
C) Freedom from hydrogen embrittlement.
D) Room temperature process - no molten metal baths or noxious fumes.
E) Elimination of 'stickers'. *In the molten metal/hot dip process, parts often stick together requiring (in some cases) 100 percent inspection after processing.*
F) Simple waste treatment.
G) No re-tap after plating. *In the molten-metal dip process, internal threaded components must either be plated and re-tapped to clean out thread roots, or plated as blanks and then thread tapped. Neither method provides corrosion protection for threaded areas. With the Mechanical Galvanizing process, normally overtapped internal threaded components are plated per standard procedure. Plating protection is provided in all recessed areas, and there are no problems with ease of assembly or thread fit.*
H) Low energy requirements/attractive economics.
I) Ability to deposit multi-metal or alloy coatings for increased corrosion protection. *In salt spray testing, the addition of 25 percent tin metal powder to the heavy zinc coating (0.021 in. or 0.053mm) provides up to 400 additional hours of corrosion protection (Table 1).*
J) Smoother and improved appearance over the molten metal dip coating application.

USERS AND SPECIFIERS:

Because Mechanical Plating is a unique process for applying corrosion-protective coatings, the list of its users is quite extensive. Primary users of the process are industries concerned with these factors:

1) Hydrogen embrittlement.
2) Alloy coatings.
3) Torque drive relationships.
4) Powder metallurgy.
5) Heavy coatings.

The performance of Mechanical Plating in areas of corrosion protection can be specifically tailored to end-user requirements. Salt-spray data are available which details various finishes lasting from 24 to 20,000 hours of exposure time (*Table 2*). The performance of Mechanical Plating in areas of torque drive relationships can also be tailored to meet specific requirements (*Table 3a and 3b*).

DISADVANTAGES:

Mechanical plating technology possesses the following limitations:

TABLE 1 -- CORROSION RESISTANCE OF MECHANICALLY GALVANIZED COATINGS
SALT SPRAY TEST RESULTS - ASTM B117

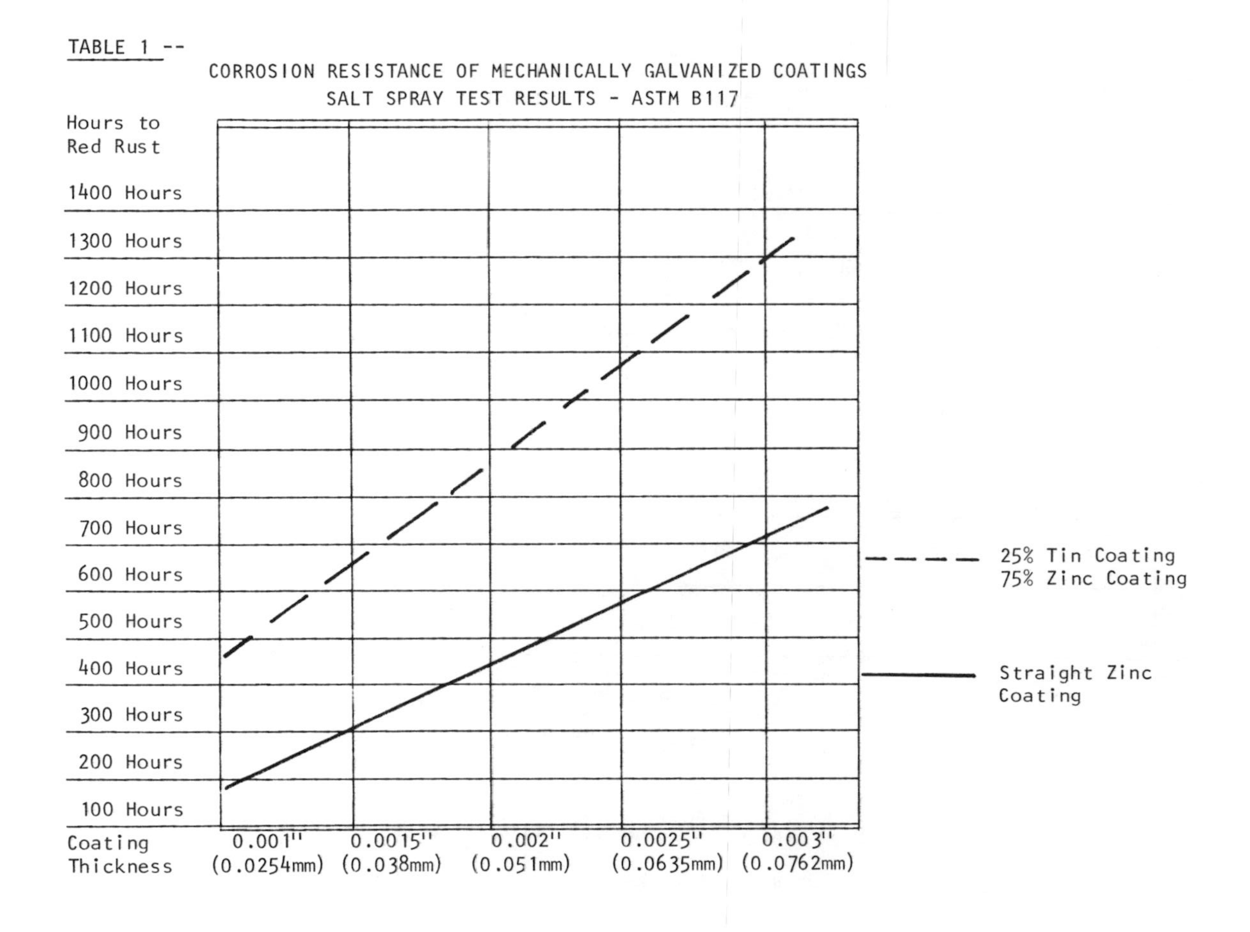

TABLE 2 - CORROSION RESISTANCE OF MECHANICALLY PLATED COATINGS

SALT SPRAY TEST RESULTS, A. S. T. M. B117 - 62

TIME IN HOURS TO RED RUST

TYPE OF COATING	THICKNESS, MILS					
	0.2	0.25	0.3	0.4	0.5	1.0
PLAIN ZINC	36		48		96	192
ZINC + CLEAR CHROMATE	36		48		96	192
ZINC + COLORED CHROMATE	108-336		120-384		168-480	264-650
PLAIN CADMIUM	50		120	200	320	
CADMIUM + CHROMATE	200-500		500-2000		2,000-20,000	
CADMIUM-TIN (50/50)	50		120	200	320	
CADMIUM-TIN (50/50) + CHROMATE	200-500		500-2000		2,000-20,000	

TABLE 3a --

Bolt Tension 16
at 28 lb/ft

3/8"-16x2", Grade 5 Clamp Loads Table 4,700 lb. min. 5,200 lb. max.	Method of Measurement SKIDMORE-WILHELM & Beam-Type Torque-Wrench	
	$\bar{X}$	$\bar{R}$
Phosphate and Oil	5820	430
Zinc, Mechanical Plating	1988	450
Zinc, MP and Chromate	2266	310
+3M Brand Torque & Tension		
Vol. %		
0	2266	310
1	4759	1445
2	5090	1000
3	4984	1140
4	5533	1200
5	6037	1350
6.25	5480	1140
7.4	5670	980
8	5880	1290
9.1	6130	1240
10	6120	1025
20	6373	1385
40	6232	1455
60	6293	680
80	6302	745
100	6557	380

TABLE 3b --

Bolt Tension 16 at 28 lb/ft	Method of Measurement	
3/8"-16x2", Grade 5 Clamp Loads Table 4,700 lb. min. 5,200 lb. max.	LEBOW FORCE SENSOR & Torque Transducer w/ Brush Recorder at 10 R.P.M.	
	$\bar{X}$	$\bar{R}$
Phosphate and Oil	4300	250
Zinc, Mechanical Plating		
Zinc, MP and Chromate	2590	300
+3M Brand Torque & Tension		
Vol. %		
0	2590	300
1	4820	220
2	4900	300
3		
4		
5	5300	660
6.25	5430	150
7.4	5430	150
8	5400	450
9.1	5190	450
10	5340	990
20		
40		
60		
80		
100		

1) Inability of present equipment to handle and plate all larger part types.
2) Inability to provide a cosmetic or micro-inch finish sometimes required.
3) Inability to deposit coatings of the less-ductile metals.

These areas are being closely scrutinized with improvements to follow as the technology expands.

RESULTS AND DISCUSSION:

The Mechanical Plating process has the unique capability of producing multi-metal finishes by co-depositing a mixture of metal powders. Many factors must be put into perspective when selecting a finish for a given application. To meet end user needs, requirements, and demands, a considerable amount of time and effort has been devoted to developing multiple metal coatings. Examples are the Cadmium/Tin coatings used widely throughout the Automotive and Marine industry. They were initially developed as alternatives to straight Cadmium coatings, and have several inherent and desirable characteristics. Straight Cadmium coatings, when given a post-plate chromate for increased corrosion protection, take on an iridescent or straw-colored appearance. However, Cadmium/Tin coatings, when given the same iridescent chromate treatment, remain silver-colored but take on the corrosion protection provided by the post-plate treatment. The Tin does not accept the iridescent color but the overall coating does exhibit increased corrosion protection. Results reveal a finish that has the corrosion protection equal to or better than straight Cadmium, that remains silver where brighter coatings are called for, has similar torque drive relationships to straight Cadmium, and is cheaper than straight Cadmium to produce. Market prices of metal powders fluctuate at a very uncertain and unpredictable rate. The cost of Mechanical Plating is directly attributable to the ups and downs of metal powder prices.

An additional and very important benefit is a 50 percent reduction in Cadmium consumption. While continuing to maintain the integrity of the protective finish the overall Cadmium usage has been significantly reduced, the end user does not suffer product deficiency, and the loss of Cadmium into the environment is minimized.

Automotive engineers are constantly adapting lighter metals for weight reduction, subsequently reducing gas consumption and pollution. Aluminum is successfully replacing steel in many applications, and continues to be looked at favorable as a viable means to further help the industry produce lighter weight vehicles.

Cadmium/Tin coated fasteners have proven superior to Zinc coated fasteners in corrosion tests against Aluminum. The Cadmium/Tin coatings last longer and have minimal destructive electrolytic action with Aluminum.

After 1000-hour salt-spray test, bolt-washer assemblies plated with Zinc with yellow chromate then fastened in chromated 6061 Aluminum channel, were severely corroded and the Aluminum was pitted due to galvanic corrosion. Similar bolt-washer assemblies mechanically plated with a 50/50 Cadmium/Tin combination with a yellow chromate revealed virtually no corrosion. The 6061 chromated Aluminum channel was also essentially corrosion-free (*See Figures 4a and 4b*).

Another application for Cadmium/Tin coatings within the transportation industry is for fasteners used in the construction of over-the-road tractors and trailers. Various mechanically plated Cadmium/Tin fasteners are used to attach and secure Aluminum, Aluminum to Steel, and Aluminum to Plywood. Subject to destructive environments in all types of roadway circumstances, these fasteners are an excellent demonstration of the coatings effectiveness.

Cadmium/Zinc is yet another co-deposit coating that is proving very promising in the metal finishing field. This finish, still undergoing rigid testing, was primarily developed as a high performance coating for components that fasten dissimilar or bimetallic surfaces. The trenchant posture of this coating is the fact that up to 75 percent of Cadmium can be eliminated from a given procedure with no deterioration in end use affects. Plating costs to produce the Cadmium/Zinc coatings are significantly less than straight Cadmium or Cadmium/Tin coatings.

Material costs only for the three finishes are as follows:

Cadmium/Zinc - $6.06/100 lbs. (45.3kg)
Straight Cadmium - $11.56/100 lbs. (45.3kg)
Cadmium/Tin - $12.18/100 lbs. (45.3kg)

Cost figures for U-Clip fasteners plated to 0.00025" (0.00635mm) at 64 lbs./cubic foot (28.99kg/0.028 cu.m) volume with a surface area of 150 sq.ft./100 lbs. (13.95 sq.m/45.3kg).

Table 5 will show how various mechanically plated coatings performed in salt spray environment when tested against Aluminum. Unalloyed Zinc coatings were not tested in this series since previous evaluations had displayed undesirable galvanic characteristics in contact with Aluminum. Straight Cadmium coatings were not tested because of Government and Industry concern over its toxicity. Therefore, alternative plating methods were considered for reduction of total Cadmium consumption and subsequent discharge to the environment.

An ongoing outdoor corrosion study is currently being conducted at various different sites including Kure Beach, N.C. - 80 and 800 ft. from the ocean (salt water atmosphere), Kearney, N.J. - heavy industrial environment, Detroit, Mich. - urban or mild industrial environment, and Sloatsburg (Sterling Forest), N.Y. - rural environment. Over 8,000

Fig. 4a - Zinc mechanically plated bolt/ washer assemblies reveal severe corrosion damage after 1000 hr. salt spray test. The Aluminum channel (6061) also reveals corrosion damage.

Fig. 4b - Cadmium/Tin mechanically plated bolt/washer assemblies reveal virtually no corrosion after 1000 hr. salt spray test. The Aluminum channel (6061) also is virtually corrosion free.

TABLE 5 --

COATING THICKNESS	TYPE OF COATING	TYPE OF ALUMINUM	FASTENER HOURS TO RED RUST	FIRST NOTED EXFOLIATION OR PITTING DAMAGE ON AL PANELS
0.00025'' (0.0063mm)	25% Cad 75% Zinc	Plain 6061	2160 Hours	1512 Hours
0.0005'' (0.0127mm)	25% Cad 75% Zinc	Plain 6061	No Red Rust at 6720 Hours	1512 Hours
0.00025'' (0.0063mm)	50% Cad 50% Zinc	Plain 6061	5904 Hours	1512 Hours
0.0005'' (0.0127mm)	50% Cad 50% Zinc	Plain 6061	No Red Rust at 6720 Hours	1512 Hours
0.00025'' (0.0063mm)	25% Cad 75% Zinc	Buffed 7029	No Red Rust at 2064 Hours	No visible damage at 2064 Hours
0.0005'' (0.0127mm)	50% Cad 50% Tin	Plain 6061	1360 Hours	1080 Hours
0.0005'' (0.0127mm)	25% Zinc 75% Solder	Plain 6061	1656 Hours	1176 Hours

ALL U-CLIP FASTENERS TESTED RECEIVED A CHROMATE DIP TREATMENT AFTER PLATING AND WERE FIRMLY ATTACHED TO THE ALUMINUM PANELS.

mechanically plated flat washers of various thicknesses and finishes have been under test for over 2½ years. In all locations Cadmium/Zinc combinations are out performing all other tested samples.

SUMMARY:

Mechanical Plating utilizes mechanical energy to provide parts that can be batch-handled with various protective finishes. The part types, along with selected glass bead impact media and finely divided metal powder, are barrel-tumbled in a chemically controlled aqueous environment. The metal powder added is cold-welded to the part types which provides the parts with an adherent, sacrificially protective metallic coating.

Mechanical plating has the unique capability of enabling deposition of multiple metals without inherent process deterrents. It is currently being utilized in replacement of a percentage of Cadmium by alternative metals having their own special advantages. This enables retention of the Cadmium characteristics which have proven advantageous while at the same time reducing the total amount of Cadmium consumed for any given quantity of parts plated.

Mechanical plating provides the end-user with product reliability, freedom from hydrogen embrittlement, consistent corrosion protection, uniform controlled coatings, proper torque drive relationships, tailor-made coatings for specific corrosion and bimetallic fastening situations, and competitive economics.

CORROSION - DECORATIVE COATINGS FOR ALUMINUM

Robert J. Leipertz
Reynolds Metals Company

INTRODUCTION

High purity aluminum has a relatively high degree of corrosion resistance and needs less protection than most metals. However, pure aluminum is soft and has low tensile and yield strength. To increase these mechanical properties of tensile and yield, certain alloying elements are added. These additions, in turn, modify the corrosion resistance of the aluminum. To overcome this problem, a number of protective coatings or surface treatments can be used.

The wide variety of interests and technical skills represented by the audience makes it difficult to be specific in discussing protective coatings for aluminum. Aluminum is used in a number of industries and market areas. I will attempt to give a brief overview of the industries and some of the protective systems used. In some cases, coatings are used for corrosion protection as well as for aesthetic reasons. It is difficult to tell which is the primary reason for a coating. Many times, protective coatings are applied to pass laboratory tests that may or may not reflect actual service conditions.

COATINGS

Coatings as used in this presentation are materials that are deposited on, or react with, aluminum to form a distinct protective barrier on the surface of the metal. Methods of coating applications will not be discussed. Only the final film in its broad generic terminology will be covered. The ideal protective coating must have the following properties.

1. It should be continuous, or substantially so, and should be impervious to gasses and liquids.

2. It should be inert or almost insoluble in its environment. It must, therefore, possess low solubility in water, acids, and alkalis.

3. It should not electrolytically accelerate the attack on the base metal.

4. It should be resistant to mechanical injury such as abrasion or scratching. If the coating is weak or thin, it should be self-repairing.

5. It should bond readily with paints and other organic finishing materials.

Some of the protective coatings that will be considered are:

Electrochemical
Chemical Reactive
Organic - Paints, Sanitary

Some of the aforementioned coatings may be used by themselves or, in some cases, in combination with one of the other coatings. Typical uses of these coatings will be discussed when markets and products are reviewed.

Protective coatings, whether they be for decorative or corrosion resistance, must be applied to a thoroughly cleaned metal surface.

The use of conversion or chemical reactive coatings is the best method for preparing aluminum surfaces prior to their being painted. Chrome-oxide type coatings by themselves have excellent corrosion resistance. They are soft and have very little abrasion resistance. This type coating is used with military equipment. The chrome-phosphate type conversion coatings provide excellent paint bond, but by themselves do not have quite the corrosion resistance of the chrome-oxide type.

REASONS

Let us take a look at some reasons for using protective coatings for corrosion resistance.

Economics
Extended Product Life
Materials Advantage
Product Protection
Aesthetic

Material costs certainly play a key role in selection of a metal. However, because of the environment, some protective coating is needed to make a viable product.

The second reason, extended life, needs no elaboration. Today's consumer protection laws require that the total product must last as long as normal life expectancy of that product.

Materials advantage could be the heat conductivity, reflectivity, or the material lightness of the metal, but these properties by themselves will not withstand the environment to which the product is exposed.

In packaging applications, the food product must be protected from the effects of the container on the food. In soft drinks that have been colored to identify with certain fruit flavors, reactions will sometimes occur that render the soft drink completely colorless. Imagine picking up a glass of colorless liquid that smells and tastes like orange or grape, but looks like water.

Aluminum storm doors and storm windows have been made for a number of years with a "natural finish," i.e., mill finish or the "as-rolled" surface. Anodic finishes (electrochemical) have been and are still being used to increase the corrosion resistance. The homeowner now wants these products to match the trim on his house. The primary reason for painting in this instance is for aesthetic reasons, but a secondary benefit is corrosion prevention. Storm doors and windows with the natural finish will show corrosion after a number of years. The length of time will depend on the atmosphere. Corrosion will occur in seacoast areas and highly polluted atmospheres in less time than it will take place in rural climates.

INDUSTRIES

We need to look at the various industries where protective coatings are applied to aluminum and at some of the general protective coatings used. The table included in this paper shows the different protective systems employed in various market/product areas. The present aluminum production capacity is over 8 billion pounds annually.

Industries which I shall cover, and the relative percentage of aluminum used in each, are as follows:

Transportation	- 21%
Containers	- 21%
Construction	- 23%
Consumer Durables	- 8%

The balance of the aluminum industry's production, 27%, is consumed by the electrical, machinery, export, and miscellaneous market areas.

PROTECTIVE COATINGS

INDUSTRY	MARKET	HOW USED	COATING CLASS	COATING TYPE
Transportation	Marine - Sea	Below water line	Organic - Paint	Epoxy or Vinyl Primer + Vinyl Topcoat & Anti-Fouling (TBTA or TBTO)
	Marine - Sea	Super Structure	Organic - Paint	Wash Primer MIL-P-15328C + Topcoat--Acrylic, Alkyd, Vinyl, etc.
	Highway	Truck Body and Frames Trailer Panels	Organic - Paint	Conversion treatment plus Epoxy or Alkyd Prime - Topcoat Acrylic, Urethane and Polyesters
	Automotive	Bumpers & Trim Bumpers	Electrochemical or Plated	Bright Dip & Anodize or Bright Chromium Plate
	Automotive	Sheet Metal Wheels	Organic-Paint Organic-Paint	Phosphate treatment - Prime + Topcoat Clear Acrylic or Urethane
	Aircraft	Structural Skin	Chemical Organic-Paint	Conversion Coatings - Chrome-Oxide Epoxy - Urethanes
Containers	Beer & Beverage	Interior - Body & Can Ends	Organic-Sanitary	Epoxy, Epoxy Modified with Vinyl
	Potted Meats Fish	Interior - Exterior Interior - Exterior	Organic-Sanitary	Vinyl Epoxy, Epoxy, Epoxy Phenolic - Epoxy
	Snack Foods	Bodies & Ends	Organic-Sanitary	Same as Potted Meats
Construction	Industrial Bldg.	Skin	Organic-Paint	Polyvinyldene Fluoride, Siliconized Acrylics or Polyesters
	Architectural Bldg.	Skins-Extrusions	Electrochemical	Integral Color Anodizing
	Residential	Siding, Soffits, Trim	Organic-Paint	Acrylics or PVC Plastisols
	Residential	Windows & Doors Windows & Doors	Electrochemical Organic-Paint	Anodic Acrylics or Polyesters
	Residential	Shingles	Organic-Paint	PVF, Acrylic, Siliconized Acrylic or Polyester
Consumer Durables	Appliances Refrigerators	Liners-Shelving	Organic-Paint	Vinyls, Polyesters, Epoxy, Alkyd Backer
		Evaporators	Electrochemical	Anodize - Clear
	Furniture	Lawn & Garden	Organic-Paint, Electrochemical	Acrylic, Polyester Anodic with Color
	Solar	Swimming Pool	Organic-Paint	Siliconized Polyester, PVF

TRANSPORTATION

The transportation industry utilized 21% of the aluminum shipments in 1977. Total aluminum usage in 1977 as reported by The Aluminum Association was 1.5 billion pounds for automotive uses. An additional 1.4 billion pounds were used in trucks, buses, aerospace, and other transportation fields. In the automotive market area, protective coatings are required for protection against the corrosive action of road salts, radiator cooling waters, and airborne contaminants. On bright metal trim such as bumpers, headlight bezels, and grilles, anodic (electrochemical) coatings are generally used. A more recent development of bright aluminum wheels utilizes clear urethane or acrylic paints. Automotive body sheet metal is presently given a zinc phosphate metal treatment, primed, and painted according to standard practices employed by the car manufacturer. Highway truck tractor rail frames are most often given a chromium conversion coating followed by an epoxy primer or an alkyd primer, depending on the manufacturer's choice. Tractor cabs made of aluminum are cleaned, conversion coated, primed, and finished with either acrylic or urethane modified paints.

CONTAINERS AND PACKAGING

Any coating systems used in contact with food products must meet the requirements of the Federal Drug Administration (FDA) or the United States Department of Agriculture (USDA). Metal pick-up of only several parts per million can result in an unusable food product, either from a flavor or from an appearance standpoint. This industry accounts for 21% of the aluminum poundage produced. Slightly over 25 billion cans were produced in 1977, over one million tons. Beer and beverage containers are coated with epoxy, or epoxy modified with vinyl. Shallow drawn containers such as potted meat cans and snack food cans are protected with modified vinyl and also epoxy coatings. Epoxy phenolic coatings are generally used in the fish packaging industry. Container metal is always coated on both sides. The inside is coated for product protection and the outside to prevent water staining during the processing cycle and to aid in the can manufacturing process.

CONSTRUCTION

The construction industry accounts for 23% of the aluminum production. Coatings are primarily used for aesthetic reasons, but in some cases industrial buildings require coatings for corrosion protection. Industrial buildings are usually coated with polyvinyldene fluoride (PVF), vinyl plastisols or vinyl organosols. Commercial buildings have finishes which utilize siliconized polyester or acrylic paints. These finishes are used on metal skin, mansard roofs, and shingles. High-rise architectural buildings are anodized with an integral color finish. Residential siding, guttering, downspouts, soffets, and trim accessories may be coated with either acrylics, polyesters, or vinyl coatings, depending on the manufacturer's preference.

CONSUMER DURABLES

This industry accounts for 8% of the aluminum produced in 1977. Aluminum is used as refrigerator liners, evaporators, shelves, air-conditioning shrouds, solar heaters, and numerous other applications. For these products, vinyls, polyesters, and alkyd paints are used predominantly.

I have attempted to give a quick summary of the protective coatings used in the aluminum industry. There are a number of coatings which I have not mentioned. Those covered here are the most widely used protective systems in the aluminum industry.

REFERENCES

Information on the properties of protective coatings was taken from the Wernick & Pinner publication, "Surface Treatment of Aluminum." The production and relative consumption figures for the various industries were provided by the Aluminum Association.

A DIFFUSED NICKEL BOND WITH A POST PLATE

James F. Braden
Continental Screw Company

This paper deals with a diffusion bonded nickel finish which with the addition of cadmium plus a variety of chromate and dichromate overlays offers excellent resistance to rust and corrosion under extreme conditions. The published claims for this process are 2,000 hours in the standard neutral salt spray test without red rust. However, many parts have gone in excess of 4,000 hours in the salt spray with no red rust.

There are several parts being tested now on a roof rack in New Bedford, Massachusetts. These parts have been exposed to the elements over 4,500 hours as of September 14, 1978 with no signs of red rust. The parts being tested are mostly thread forming screws. The tests are being conducted in actual applications. The fasteners are seated with conventional power tools. In this way we can determine how effective the finish is in resisting rust and corrosion as a result of damage to the wrenching surfaces. We are also able to determine the effect of the type of material being fastened. The fastening sites that have been used range from steel, painted steel, aluminum extrusions, cast aluminum and stainless steel.

Table 1. Neutral Salt Spray Test per ASTM B117

#8 POZIDRIV®Pan Head Steel Self Drilling Screws with SANBOND (tm), Cadmium and Clear Chromate	3,000 hours <u>NO</u> Red Rust
5/16-18 Hex Washer Head Steel TAPTITE® Screws standard Cadmium .00015" minimum thickness	72 hours Red Rust
M12 x 1.75 TORX® Pan Head Steel TAPTITE Screws Polymer Phosphate	48 hours Red Rust
#8 POZIDRIV Pan Head Steel Self Drilling Screws Multi-layer Chromium	168 hours Red Rust

5/16-18 Hex Washer Head Steel TAPTITE Screw Phosphate and Oil	140 hours Red Rust
5/16-18 Hex Washer Head Steel TAPTITE Screw Metallic Lacquer Coating	144 hours Red Rust
#8 POZIDRIV Pan Head Steel Self Drilling Screw Mechanical Cad-Tin	168 hours Red Rust
5/16-18 Hex Washer Head Steel TAPTITE Screw Lubricated Resin Coating	72 hours Red Rust

Metallographic mounts made of parts that have been finished with this diffusion process show that there is an actual alloying of the nickel with the base metal. It can be shown on a metallographic mount that the thickness of the nickel layer appears to be thicker due to the alloying with the base metal. This apparent increase in the thickness of the nickel layer has been measured and found to be 30% to 40% greater than before heat treating.

The nickel plating is not the standard lamina crystaline structure used for decorative purposes. It is more of a cubic crystaline structure which accounts for its "throwing power." This also makes it possible for it to be more readily diffused than the standard nickel plating. There are several beneficial effects of this alloying of the nickel and the base metal. First, it provides an even and complete coverage of irregular surfaces. Second, it demonstrates excellent throwing ability - that is, it fills crevices and even microscopic imperfections. This attribute is particularly important because many times the rust and corrosion start at these imperfections. Third, this bonded nickel surface provides a beneficial barrier which, while permeable to carbon, has a favorable throttling effect in controlling excessive carburization. This is a favorable characteristic when the parts are case hardened. It is also beneficial in that this nickel layer tends to balance the surface carbon content when neutral hardened parts are being considered.

Following the heat treatment of the SANBOND Nickel coated parts, a special cleaning operation (without the use of acid) is performed in preparation for the application of the cadmium overplate. This cleaning without the use of acid is important in that one of the most common sources of hydrogen is acid cleaning. The fact that the part is heat treated after the initial nickel plating drives any hydrogen out of the parts. The use of this process virtually eliminates the problem of hydrogen embrittlement. This diffusion nickel bonded coating is particularly well suited for carburized products. It also can be used to great advantage in coating springs, clips and similar parts that are very susceptible to hydrogen embrittlement.

It is noteworthy that this finish does not have any detrimental effects on the performance of carburized products such as high performance thread forming screws and self drilling screws. Tests have been conducted using a self drilling screw with the diffused bonded nickel plus a bright cadmium overplate. The results show that this finish combination has superior corrosion resistance. Also, it can be shown that the drill times are well below industry specifications for both chrome and conventional finishes.

Table II. Drill Times for
#8-18 x .43 POZIDRIV Round Washer Head
Self Drilling Screws with SANBOND-Cadmium Finish

Standard .062" Steel Test Plate

25 samples

Load	45 lbs.*	30 lbs.**
Drill Time Average	.663 seconds	1.433 seconds
Range	.47-.88 seconds	.84-2.23 seconds
Standard Deviation	.102 seconds	.348 seconds

Specification - 3 seconds maximum in each case

*Standard Load for Chromium plated product

**Standard Load for conventional finishes

The application of this coating to high performance thread forming screws such as the TAPTITE Thread Forming Screw makes it possible to expand the use of this type of fastener. The cost savings possible by using these fasteners can now be appreciated. Historically, there were applications for threaded fasteners that had rust and corrosion resistant requirements that made it necessary to use stainless steel. When the stainless steel used was the austenitic type it was usually not possible to use a thread forming screw. The reason for this is the austenitic type stainless steel cannot be heat treated, hence, the fastener can not be made harder than the nut member. The excellent rust and corrosion characteristics of this SANBOND finish combined with the fact that we can carburize through the initial nickel layer enables us to consider a compromise in material and realize substantial cost savings.

It was written recently that metallic corrosion may be costing the United States about 70 billion dollars each year.(1) These sort of cost figures are certainly sufficient motivation to find better ways to reduce the rust and corrosion of metallic parts. This diffused nickel bond coating with a cadmium post plate would certainly appear to be a good solution to some of the corrosion and rust problems. It is all the more interesting when applied to the cost savings type of product that have been mentioned i.e., thread forming screws and self drilling screws. It certainly is applicable to other types of products.

One other interesting aspect of this finish is the fact that heat treatment is involved. For instance, a fastener can be heat treated to a higher strength level than an austenitic stainless steel fastener of the same size. We now have a fastener with a higher strength and nearly as good rust and corrosion resistance as the stainless steel part. This thought can be taken one step farther by considering reducing the size of the fastener which is now possible because of the increased strength. The obvious advantages here are the reduced weight which is particularly interesting to the automotive manufacturer, for instance. The automotive industry is not the only industry that is becoming more and more weight conscious, however.

The virtual elimination of hydrogen embrittlement is a particularly important attribute. Hydrogen embrittlement is a much misunderstood phenomena. Many delayed failures in carburized products, for instance, are contributed mistakenly to hydrogen embrittlement. There are delayed failures due to other reasons that are unrelated to the classic hydrogen failures.

It was mentioned earlier that any hydrogen that was introduced during the nickel plating would be driven off during the subsequent heat treatment. Also, there have been numerous studies done concerning the amount of hydrogen that is diffused through various materials, one of these materials being nickel. The consensus of these studies is that a layer(2) of nickel substantially retards the rate of diffusion of hydrogen. There is every reason to believe that we can now post plate these materials without danger of introducing free hydrogen; particularly, if acid is not used in the cleaning operation.

The elimination of the possibility of hydrogen now further expands potential applications for the finish. The product that comes to mind is the neutral hardened high strength bolt. Historically, many people would not consider electro plated high strength bolts because of the potential hydrogen hazard. Substantial cost savings, improved product quality and reliability can be appreciated if the use of this finish is extended to high strength bolts.

Conclusions: The diffused nickel bond with a post plate of cadmium has excellent rust and corrosion resistant characteristics. The bonding or alloying of the nickel with the base metal provides an even undercoating. The surface imperfections in the part are filled and not bridged over by the diffusion of the alloy layer. This enhances the corrosion resistance of the part because very often the corrosion starts at one of the imperfections. The diffusion bond of nickel also improves the resistance of corrosion and rust due to wrench damage to the wrenching surfaces of threaded fasteners.

This finish enables us to expand the use of cost saving fasteners such as thread forming screws and self drilling screws. It also enables us to consider inexpensive substitutes for stainless steel parts. These types of applications concerning the stainless steel parts can be further enhanced when we consider that it is possible to reduce the size of the part because the plain carbon steel heat treated part can be made stronger.

A comparison test of a TAPTITE Thread Forming Screw with three different finishes in an actual application illustrates the superior performance of the diffused nickel bond finish.

CORROSION TEST OF 10-24 x .75 Indented Hex Head TAPTITE Screws in aluminum die casting

Reason for test: To evaluate the corrosion action in neutral salt spray on several finishes on fasteners in an aluminum casting used for exterior electrical connections. TAPTITE Screws of stainless steel and colored manganese phosphate were compared to steel TAPTITE Screws with SANBOND-Cadmium finish.

Samples: 10-24 Hex Head TAPTITE Screws

A. SANBOND-Cadmium
B. 305 series stainless steel
C. Colored manganese phosphate

Anodized and plain aluminum castings were assembled with TAPTITE Screws with each of the above coatings. The screws were used to attach the cover plate to the cast housing.

Unassembled screws with the above coatings were also tested along with screws with plain zinc and metallic lacquer finishes.

Background: Reportedly in outdoor service, the 305 stainless steel TAPTITE Screws would "seize and gall" upon removal.

Results: The 500 hour salt spray exposure of the assembly clearly demonstrated the superiority of the SANBOND-Cadmium Finish on fasteners in this aluminum casting.

Alone all finishes except the SANBOND developed rust ----

305 stainless steel in 144 hours
colored manganese phosphate in 48 hours
bright zinc in 32 hours

SANBOND-CADMIUM no rust

In the casting the stainless steel exhibited the binding action similar to the field complaint. Galvanic corrosion of the aluminum by the 305 stainless steel caused the binding; most notable in the threads and around the screws of the anodized casting.

Contrastingly, the SANBOND-Cadmium exhibited no galvanic action - no binding. Further, the anodized casting with SANBOND-CADMIUM TAPTITE Screws showed the least casting corrosion. The rusted colored manganese phosphate showed the most casting corrosion

The binding action could be measured in part by the removal torque. The SANBOND-Cadmium had the lowest removal torques in comparison to the stainless steel TAPTITE Screws (normal prevailing torque still present). The binding effect can quantitatively be seen in the torque removal in the data.

A. Data - Salt Spray - Screws Along

Sample		Observations
1. SANBOND-Cadmium 10-24 Hex Head Steel TAPTITE Screws	(A)	1200 hours, no rust, gray, still in test, 9/27/77 (10 samples)
	(B)	500 hours, no rust, light gray, tested with next two sets (10 samples) test terminated without failure

2. 300 grade stainless steel 10-24 Hex Head TAPTITE Screw (AISI 384 or 305)

   144 hours spot of rust st rted 4 of 10

   500 hours slight rust on heads of 4, all slight rust in threads test terminated

3. Colored manganese phosphate TAPTITE Screws

   48 hours rust on all 10 and rust continued to increase

   500 hours 100% heavy crusted rust when test was terminated

4. Bright zinc 10-32 x .35 Phillips Pan Head Steel Screw

   36 hours white corrosion rust started, red rust 72-96 hours total.

   500 hours heavily crusted rust

5. Metallic Lacquer Coated 1/4-20 Steel Bolt

   500 hours on 1/8" diameter spot of rust, several smaller spots dull gray with film of white

B. Data - Salt Spray in Assembly

500 hours, 10-24 screws

Sample Finish	Anodized Aluminum	Not Anodized Aluminum
SANBOND-Cadmium	Screws: light gray-no rust Casting: starting white corrosion, very minimal & not associated with SANBOND-Cadmium fasteners	Screws: gray - no rust with run-over corrosion from casting, casting voluminous white corrosion
Stainless Steel	Screws: dulled slightly-no rust casting 10%-15% white, notable near screws, evident galvanic action inside threads	Screws: slightly dulled, no rust Casting: same as above.

Sample Finish	Anodized Aluminum	Not Anodized Aluminum
Colored Manganese Phosphate	Screws: 2 - rusted 1 - blackened 50% white corrosion, also near screws	Screws: All 3 rusted Casting: voluminous white corrosion

C. Data - 10-24 TAPTITE Screws - Cover Plate Removal Torque-After 500 hours Salt Spray In Assembly

Finish on Screws	Anodized Casting Lb-In-Off	Not Anodized Casting Lb-In-Off	Average
SANBOND Cadmium Dicromate on Steel x .75 long	16.1 14.9 16.7	26.3 17.9 18.4	$\overline{X}$ = 18.38 Sigma = 4.08
Retighten to 25	OK	OK	
Stainless Steel x .75 long	16.4 11.2 16.6	23.3 37.9 22.5	$\overline{X}$ = 21.32 Sigma = 9.26
Retighten to 25	OK	OK	
Colored Manganese Phosphate on steel x .6 long	11.1 12.3 Rust 23.3 Rust	19.8 Rust 24.1 Rust 20.0 Rust	$\overline{X}$ = 18.43 Sigma = 5.50
Retighten to 25	Stripped	2-OK, 1 Stripped one at 17	

SANBOND finish is available with decorative dichromate overlays in clear, yellow, gold, blue, olive, black and other colors.

This finish is extremely interesting and has potential to both improve rust and corrosion resistance and at the same time show substantial cost savings.

REFERENCES

(1) "Economic Study by the Commerce Department's National Bureau of Standards"

(2) Freiman, L.I. och Timov, V.A., "Reducing the electrodiffusion of hydrogen in iron and steel by aid of thin metallic coatings", Zhurnal Fiz. Khimii 30 (1956) 882-888

Matshshima, I. och Uhlig, H.H. "Protection of Steel from Hydrogen Cracking by Metallic Coatings", Proc. 3 Int. Congress Metallic Corrosion
3 (1966) 406-412

Tardif, H.P. och Marquis, H. "Protection of Steel from Hydrogen by Surface Coatings"
Canadian Metall. Quarterly 1:2 (1962) 153-171

Schuldiner, S. and Hoare, J.P. "Transport of Hydrogen through Palladium - Clad Electrodes", Can. J. Chem. 37 (1959) 228-237.

POZIDRIV is registered trademark of Phillips International.

TAPTITE is registered trademark of Research Engineering and Manufacturing, Inc. a unit of AMCA International.

TORX is a registered trademark of Camcar Textron.

SANBOND is a trademark of Rederiaktiebolaget Nordstjernam.

A NEW CONCEPT IN ALUMINUM PROTECTION

Peggy Gaul
Peter Didrichsons
Dow Corning Corporation

ABSTRACT

This paper describes the development of a new technology for protecting aluminum surfaces. Aluminum which has been treated with this material will maintain its bright surface appearance in a wide variety of corrosive and abrasive environments.

This material possesses both organic and inorganic qualities. The treatment solution handles much like a conventional organic coating, and can be applied using dip, flow, or spray coating methods. The cured film properties approach those of glass. It is clear, hard, and unaffected by a wide variety of organic solvents.

Extensive laboratory and field testing has shown that aluminum surfaces protected with very thin films (0.05-0.3 mils) of this material are highly resistant to many chemicals, corrosives, road salts, heat, UV degradation, and abrasion. Adhesion to properly cleaned aluminum is exceptional, even under such adverse environments as boiling water, high humidity, and steam.

The data illustrates the treatment's performance in automotive tests such as salt spray resistance, simulated weathering, water immersion, thermal shock, chemical and abrasion resistance. Actual field test results involving automotive, cookware, and other applications will also be discussed.

INTRODUCTION

A newly developed technology for the protection of bright aluminum overcomes several of the weaknesses of unprotected aluminum, anodized aluminum, and clear organic coatings for aluminum. The new metal surface treatment involves the creation of a thin, tightly bonded, clear, glossy, very hard film. This material is unique in that it can be handled and applied like a conventional organic paint while giving cured properties similar to an inorganic treatment. This one-part system provides both improved corrosion and abrasion resistance when compared to unprotected aluminum and aluminum protected with clear organic coatings.

ALUMINUM PROTECTION

Pure aluminum reacts with oxygen and water in the atmosphere to form a self-passivating layer of Al_2O_3. This oxide layer is so unreactive and adherent that aluminum is resistant to distilled water, most neutral solutions, many weakly acidic solutions, and to the atmosphere. But other methods of corrosion prevention are required in atmospheres contaminated with chlorides or other halogen ions, coal dust, alkaline chemical discharge or in highly corrosive and abrasive environments. Highly buffed or chemically brightened aluminum is particularly susceptible to appearance degradation and corrosion. This "chrome-like" aluminum surface will quickly dull, discolor, and lose its reflectivity if left unprotected.

The most commonly used corrosion control techniques for bright aluminum surfaces are anodizing and organic coatings. Each of these has its limitations. Anodized aluminum is hard and resists corrosion but anodizing is an energy intensive, difficult electrochemical process that can dull bright aluminum. Anodizing is not always an option since not all aluminum alloys can be clear anodized to an acceptable appearance. Also, like the unprotected metal, anodized aluminum is susceptible to alkaline attack. Clear organic coatings are an economically attractive choice for corrosion protection since they are easily handled and applied. These organics cannot be used in all environments, however, since many "yellow" on long term exterior exposure and have poor abrasion and water resistance.

NEW MATERIAL

DOW CORNING® Metal Surface Treatment (MST) is soluble in low molecular weight alcohols and ether alcohols. The solvent system in commercial use is non-red label (rated combustible rather than flammable). All solvents in the system are Rule 66 exempt, which minimizes pollution problems. This alcoholic system has a low viscosity (8-12 centipoises) and thus this material can be applied like conventional organic coatings. Dip, flow or spray coating methods may all be used. Conventional paint application equipment may be utilized.

This material is a thermoset coating. After application and a short solvent flash off period, a heat cure is needed to accelerate the curing reaction. The time and temperature requirements are minimal. This system requires less energy than most thermosetting organic coatings.

The cured film of MST is very hard, generally 2 or 3 pencil hardnesses greater than the aluminum substrate. It is very clear and does not alter the appearance of the aluminum surface. Highly polished mirror aluminum surfaces coated with this material exhibit less than a 5% drop in total reflectance, as compared to uncoated.

PROPERTIES AND PERFORMANCE

The combination of final properties exhibited by this material and its ease of application make this technology unique in the clear coatings field. Currently available clear organic coatings may be equally easy to apply, but none possess the wide spectrum of performance properties necessary for use in maintaining bright aluminum in highly corrosive and abrasive environments. Anodizing in some instances may equal most of the material's performance, but requires a much more involved application procedure. The rest of this paper is devoted to illustrating the properties and performances of MST.

Adhesion to Aluminum

The primary mode of adhesion of this material to aluminum is thought to be hydrogen bonding of the MST to the surface aluminum oxide layer. Surface preparation of the aluminum is very important because anything which interferes with such bond formation would decrease adhesion. In order to insure maximum adhesion, the aluminum surface must be chemically cleaned just prior to coating. In our laboratories, we have found that soaking aluminum in an inhibited alkaline cleaner followed by a rinse with deionized water is quite effective. No primers, conversion coatings or other special processes are necessary.

Adhesion of the material to an aluminum surface thus prepared is excellent, as measured by the common crosshatch tape pull method (ASTM D-3359). In fact, MST remains chemically bound to the substrate under severe exposure conditions such as boiling water, high humidity, and steam -- critical conditions which challenge adhesion in practical performance. Treated panels have withstood more than 1,000 hours in a Cleveland Condensation Tester (ASTM D-2247) with no blistering, no roll back at the scribe lines, and no loss of adhesion. Treated panels have been repeatedly cycled from -70°F to +212°F with no loss of adhesion.

Hardness and Abrasion Resistance

The hardness of MST depends on the substrate to which it is applied. On 3003 aluminum alloy, the pencil hardness is 7-8H; on 5252 aluminum alloy, the pencil hardness is 5-6H. Pencil hardness was determined by standard method ASTM D-3363. This degree of hardness makes surfaces treated with MST very abrasion resistant. Test panels were abraded, using a Tabor Abraser as described in ASTM D-1044. Treated panels withstand 100 revolutions of a Tabor Abraser (500 gram weight CS-10F wheels) with minimal abrasion or appearance degradation. This is comparable to the performance of some anodizing treatments. Organic coatings typically degrade in appearance after just 10 cycles of the Tabor Abraser.

Weathering

MST has excellent weatherability. This material does not alter the appearance of the aluminum when applied, and does not yellow upon weathering. Treated panels have withstood 2,000 hours of accelerated weathering (Atlas Carbon Arc Weatherometer ASTM G-23 Type E) with no evidence of degradation. Even after the equivalent of 4-5 years of Arizona sunshine (500,000 Langleys) reflectivity was reduced less than 1/2 of 1 percent.

Corrosion Protection

Unlike clear organic coatings which are typically greater than 2 mils thick, very thin films of MST (0.05-0.3 mils) are very effective in protecting an aluminum surface from corrosion.

Many tests are used by different industries to predict corrosion resistance after long term use. Those tests which we have conducted to date show that the MST system provides excellent protection.

Salt Spray and CASS Tests. These tests are particularly important to the automotive industry in predicting how effective a coating will be in protecting bright aluminum parts (wheels, trim, etc.) from the corrosive effects of road salts. Treated aluminum panels have withstood over 1,000 hours of 2½% NaCl spray with no evidence of degradation (Military Specification 883A Method 10091). Corrosion occurs only at the scribe line with no undercutting corrosion from this line. Treated panels subjected to 24 hours of copper acidified salt spray show no degradation (ASTM B-368).

Mortar Test (ASTM D-3260). The mortar resistance of bright aluminum is particularly important in architectural application where there is a possibility of corrosion of the aluminum from the mortar used during building construction. Aluminum panels were treated and freshly prepared mortar was applied; these panels were then placed in an atmosphere controlled at 98% relative humidity and 70°F. After one week, untreated panels were 100% corroded; anodized panels were 100% corroded; MST treated panels were less than 5% corroded; MST treated anodized panels were less then 0.03% corroded. Thus, MST provides excellent protection against the alkaline mortar attack.

Chemical Resistance. Treated surfaces will withstand 30 days or more exposure to a wide variety of organic chemicals and solvents. Panels will withstand over 8 hours in boiling 3% acetic acid, and more than 24 hours in 0.01 M NaOH, 31% $CaCl_2$ solution. Treated aluminum panels have also been exposed to more than 200 home dishwasher cycles without degradation.

Applications

The above list of performance properties serves to illustrate that MST is truly unique as a clear treatment for aluminum. This treatment does not alter the appearance of highly polished aluminum and protects this "mirror-like" finish from corrosion, weathering and abrasion. This technology is applicable where there is a need to protect a polished aluminum surface.

It is desirable from a weight savings point of view to replace steel and zinc automotive parts with aluminum. In many of these potential replacement applications, a highly reflective surface appearance is needed for styling purposes, e.g., bumpers, wheels, trim. This surface needs a clear protective treatment. Anodizing is not always an option because of the alloys and cost involved. Many clear organics cannot withstand the wide variety of stresses of an automotive environment. MST will protect these reflective surfaces through all of these stresses. It will withstand the corrosive effects of road salts; the abrasive action of road use and washing; and the steam blasts, temperature fluctuations and alkaline environment of a commercial car wash.

An application in the early stages of commercialization is the use of MST in the cookware market. There are two metals extensively used, and very popular in the cookware industry. These are stainless steel and aluminum. Stainless steel cookware is very durable. Its finish is very attractive and resistant to tarnish and wear. Aluminum cookware has different advantages. Aluminum is an excellent conductor of heat, and thus cookware made of this metal uses cooking energy more efficiently than stainless steel. Sturdy heavy gauge aluminum cookware is very lightweight and easy to handle. There are high quality non-stick surfaces (e.g., Silverstone®) that adhere well to aluminum and thus simplify cookware cleanup.

Aluminum cookware is manufactured with a wide variety of exterior finishes, some of which are porcelain enamel, acrylic or polished natural aluminum. The polished aluminum looks very much like the much more expensive stainless steel cookware. This unprotected finish is not durable and, in fact, turns black after just a few dishwasher cycles. When protected with MST, aluminum cookware will retain this polished "stainless" appearance. Treated pans have withstood more than 100 cycles in an industrial dishwasher (equivalent to the normal household lifetime of a pan). Test panels have been boiled in a 5% Cascade® solution for more than 72 hours with no deterioration of the treated surface. The untreated surface was black. MST is thermally stable, and thus, unlike acrylics, can withstand stovetop use without charring or yellowing. MST treated cookware has the advantages of aluminum with a durable, appealing, polished finish.

The two applications mentioned are by no means the only potential markets Dow Corning envisions for this technology. Another example might be maintaining the reflectivity of solar reflectors.

To summarize, MST is unique technology in the area of clear protective coatings for aluminum that could potentially expand the market for polished aluminum.

CATHODIC ELECTRODEPOSITION: A NEW PAINTING TECHNOLOGY FOR AUTOMOTIVE INDUSTRY

M. MORIONDO - FIAT
G. BIANCHINI - IVI

1. INTRODUCTION

The principle of electrodeposition painting has been known since the beginning of the century and has greatly improved owing to the technological developement of the 60 years.
In car industry, and in particular in FIAT, such process started being included productively in 1965 with a tank for little parts. Nowadays we think that the electrophoretic painting plants are more than thousand with tanks as wide as containing 250000 lt. Till 1976 the electrodeposition painting practised in the world was being founded on the principle of the anodic deposition. The cathodic electrophoresis came to american market in 1974 with a tank for the production of little parts of electrical household appliances. But the main limit to its wide use in the industrial field consisted of the poor throwing power of such a material.
At present this limitation has been brilliantly overcome and the american car industries are rapidly drawing towards the generalized use on the car body of this new technology that allows clear improvements of resistance to corrosion valuable within the 20 - 60% according to the cataphoresis used.
In Europe FIAT has been following since 1976 the developement of the kinds of cataphoretic products appeared on the international markets. On November 1977 started extensive tests of production a tank containing 30000 lt. for the painting of little parts, and at the beginning of September 1978 started the cataphoretic painting process on a tank for car bodies containing 180000 lt.

1.1 Why the cathodic electrophoresis

The chemistry of the anodic and cathodic painting is well known, therefore we just illustrate synthetically in Fig. 1 the reactions occuring in the conventional process of anodic deposition and in the innovative one of cathodic deposition is that they may explain the essential characteristics of the applied films.

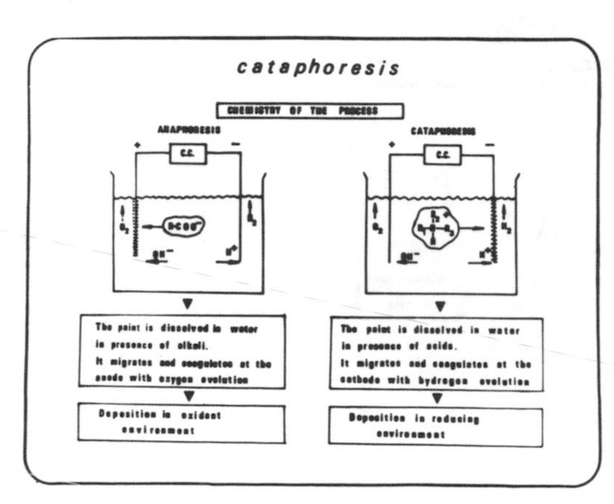

Fig. 1 Comparation between the two processes of electrodeposition in outline.

In the anodic process the reaction of the alkali on the acid polymer RCOOH insoluble in water, forms a soluble salified polymer. During the passage of continuous electric current, at the anode there is the electrolysis of water involving production of protons closely near the surface; this leads the pH to$\sim$2.

For apposite reaction an insoluble polymeric acid is formed again. The polymer coagulates on the piece connected to the anode. The anodic dissolution of the metal forms metallic ions that intervene in the coagulation of the polyelectrolyte forming insoluble metallic salts.

The same reaction occurs, when it is the matter of phosphated components, from the dissolution of the phosphate. There is oxygen evolution in form of gas, and it can react either with the polymer or with the steel sheet. In the cathodic process there is an insoluble polymeric base that, adding organic acids, becomes a soluble salt. At the cathode there is production of hydroxyls (pH$\sim$12) and the insoluble polymeric base coagulates.

The metal does not dissolve and the dissolution of zinc or iron phosphate is lower. On the contrary the present metallic ions are reduced to metal owing to electrolysis at the cathode. The hydrogen evolves in form of gas. What said above lets us expect the following qualitative advantages, reported in Table 1, shown by the cataphoresis as to the conventional anodic electrophoresis.

Table 1 Expected advantages by using a cataphoretic product

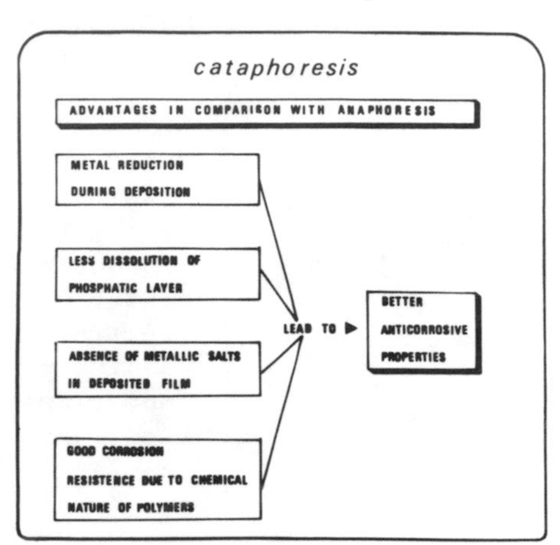

2. ELECTROCHEMICAL APPROACH (1)

2.1 Materials and methods

The resins for cataphoresis are available without pigments as a trasparent resin, and with pigments as commercial paints (black colour). The experimental equipment consists, for the electrode position of the resin, of a Pyrex double wall cylindric cell ha ving a capacity of 750 cm^3, kept at a constant temperature by means of water circulation and a rectangular cell having regu - lar parallel sides with a capacity of 1000 cm^3. For the former one we used a stainless steel plate anode surrounding the whole lateral wall of the cell (area = 300 cm^2), for the latter one a flat stainless steel plate (area = 150 cm^2). The working elec trode was, time by time, an Aluminium wire (ERBA RP 99,5%) ha - ving a diameter of 1 mm, or an Aluminium, Iron and phosphated Iron flat plate with an area of 150 cm^2.

The equipment necessary for the electrodeposition, outlined in Fig. 2, consisting of an oscilloscope with variable persistence (duration of trace after glow), a XY recorder, a coulombmeter-integrator, a potentiostat and two power supply amplifiers able to apply, at full load, 120 V and 1,2 A, allows us to analyze the electrochemical parameters; load, tension, current (this one as ohmic drop at the ends of a known resistence). Other measure

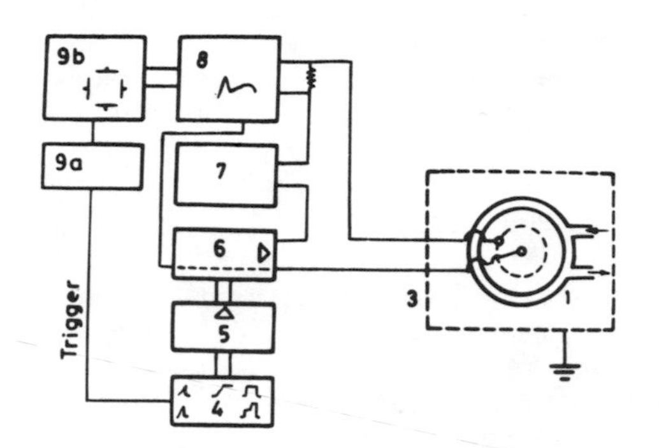

ments were obtained with an AMEL 'Metalloscan' potentiostatic apparatus an INGOLD Glass-calomel micro electrode, a Keithley (mod. 160) digital multimeter, and a Keithley (mod.370) galvanometric recorder. The analysis of the electrodeposed layer was performed utilizing a JEOL scanning electron microscope and an energy dispersive analyzer of X-rays (EDAX).

Fig. 2 Equipment for the imposition of electric field and for the survey of electrochemical parameters:
1. electrodeposition cell;-
2. to the thermostat;-
3. Faraday cage;- 4. function generator AMEL (mod.565);-
5. power pre-amplifier (mod. 551);- 6. two power amplifiers HP (mod. 6924 A);- 7. coulombmeter-integrator AMEL (mod.558)
8. XY recorder AMEL (mod.822 D)
9. and 9bis. oscilloscope HP (mod. 141 B)

2.2 Results and discussion

The cataphoresis process governed, as it generally occurs in the electrodeposition, by means of applied electric field, by electrostatic interactions on the electrode interface, by real chemical reactions and by variations of Zeta potential (2). The cathodic resins are in fact hydrophilic products in which the positive charges present in the chain are supported by Nitrogen atoms, present in aminic groups. Before being diluted to the proper concentration (15% in weight), they are neutralized with organic acids till a pH = 5,5; in such conditions the bath has a conducibility of 1,6 mS.

The application of the electric field causes at first two kinds of electrostatic interaction, reversible and irreversible, as we shall point out later and, above all, electrochemical reactions consisting of the water discharge (3), so that at the cathode we obtain:

$$2H_2O+2e^- = H_2+2OH^- \quad (A)$$

The increase of the ionic force of the solution near the electrode causes a decrease of the stability of the colloid, which tends to deposit itself, and this is due also to Zeta potential decrease; moreover the just produced OH^- ions neutralize the positive charges of the resin, so that it becomes in this way hydrophobic and is definitively deposed (3,4). In order to follow the deposition behaviour we studied the current-tension curve, obtained applying electrical fields different for intensity, wave form, cell geometry etc. The deposition during the first seconds was followed by means of the oscilloscope, the whole curve was recorded on paper. The study of the first seconds concerns above all the valuation of the electrostatic effects on the electrode: therefore we tried to determine the limits of reversibility conditions, i.e. the conditions of applied electric field with which we have the electrochemical parameters under which we do not have any effective painting. Particularly the tension under which the current remains practically and anyway zero; such a tension, which was practically null in the anaphoretical process has here a va-

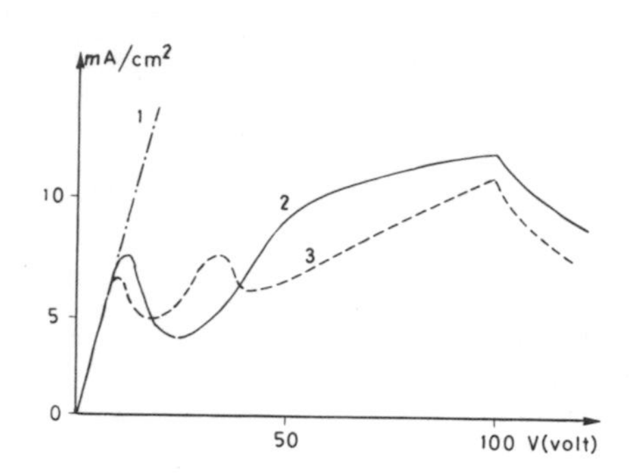

Fig. 3 Current-tension curves recorded in different mediums on anodically polarized Aluminium; Curve 1: $NaClO_4$+NaOH; pH = 8,2 Curve 2: Acrylic resin 10%; pH = 8,2 Curve 3: Acrylic resin 10%+TiO_2 Pig. pH = 8,2

lue of about 3,5 - 4V; this matter points out an initial trend towards passivity; the same behaviour we have observed in the current-tension curves before the first current peak.

We found also the maximum charges which can be applied reversibly , by applying some triangle wave tensions having increasing intensity and different scanning rates.

The effective reversibility conditions depend also, unlike the anaphoresis, from the time interval elapsed between the repeated pulses. In such conditions in fact the electrode acts as a capacitor which, having been charged, requires a certain time for the total discharge. Thus in this case there are three parameters which determine a reversible electrostatic interaction; the applied field intensity, the scanning rate, the time interval between the varius pulses. When we give a linear potential sweep we obtain a corresponding current path which is too a ramp and which sometimes includes the whole first peak of the current-tension curve (Fig. 4, curve 1). So in Table 2 we report the given tensions (ramp) with their scanning rates, and the corresponding charges and intervals in reversibility conditions.

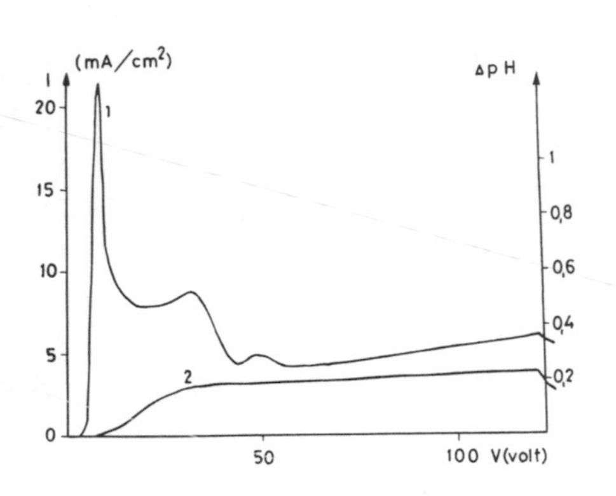

Fig. 4 Curve 1. (left axis): current-tension curve recorded on cathodically polarized Aluminium in resin for cataphoresis: 15% in weight; pH = 5,5
Curve 2. (right axis): pH path closely near the electrode during cataphoretic electrodeposition

Let us consider now the whole current-tension curve, and compare with the same one obtained by anaphoresis. In fact in this diagram (Fig. 3, curves 2 and 3) we cannot find any initial trend towards passivity, and we can see that the first peak has the same slope as the dissolution curve of the Aluminium in water solution having the same concentration and pH (Fig. 3, curve 1).

On the contrary if we measure the slope of the current-tension curve for the Aluminium cathodically polarized in water solution,

Table 2 Electrochemical data of the conditions of reversibility tension applied, scanning rate, charge applied and time interval between varius pulses

Tension applied (V)	Scanning rate (V/sec)	Charge applied $\left[\frac{\mu Coul}{cm^2}\right]$	Time [sec]
5	48	1000	~8
5	96	350	~4
10	48	5000	20*
10	192	2100	5
10	480	600	2
15	960	1000	3
20	480	3200	25*
20	960	1900	8
30	960	4500	20*

*) These data are found where the whole first peak results rever sible

this one results higher than the slope of the first peak of cata phoresis (Fig. 4, curve 1), without a real quantitative connec - tion between the two slopes. The first part of the curve is the- refore governed more by electrostatic processes than chemical ones. In Fig. 4 curve 2 we can see the path of pH variation clo sely near the cathode, according to the reaction (A) (pH max = 0,220 pH units).

3. MECHANISMS OF LAYERS GROWTH AND DISTRIBUTION OF MOLECULAR WEIGHTS ALONG THE FILM SECTION

The principal problem was to state if the layer growth takes pla ce for deposition of the following layers one on the other or, on the contrary, takes place for mechanism of growth from the bottom, that is, for penetration of the ionized polymer molecu- les through the previously deposited film and for the following discharge of them directly on the metallic substrate. We ap - proached the problem synthetizing four different alkyd resins.

Such resins were quite similar in the physico-chemical proper - ties (amine number, percentage of aromatic rings and fatty a- cids content), but are different from one another in molecular weights and chlorine content.
In Table 3 we report the characteristics of such polymers.

Table 3 Characteristics of the resins emploied for the determination of the layer growth

Resin	Molecular weight	Chlorine Content %
127	2400	0
128	2600	8,2
134	5200	0
135	5600	8,2

Chlorine atoms were introduced in resins 128 and 135 replacing phtalic anhydride with tetrachlorophtalic anhydride. Investi - gation on film growth was performed by mean of successive elec- trodeposition of different binders on the same panel.
EDAX microphotographies of sections of deposited films allowed a precise determination of the position, along the film, of chlo rine marked resins. In Fig. 5 the disposition of chlorine atoms in a film obtained with the electrodeposition resin 135 (chlori- ne containing) on 134.
In Fig. 6 the disposition of chlorine atoms in a film obtained by electrodeposition of resin 134 on 135 (chlorine containing) is shown.
As figures 5 and 6 show , one can find near the substrate the resin electrodeposited in the second stage. This means that, du ring electrodeposition, polymer molecules pass through the pre - viously deposited film and discharge directly on the metallic panel. To state different deposition rates of polymers as a fun ction of the molecular weights, two electrophoretic baths were prepared.

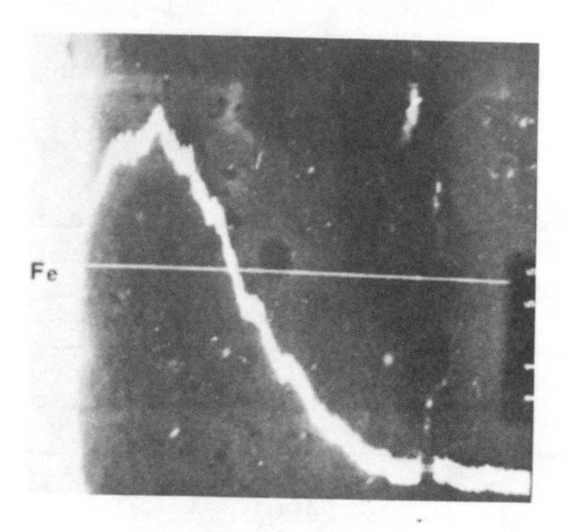

Fig. 5 Chlorine content trend along the section of the film obtained by electrodeposition of resin 135 on 134

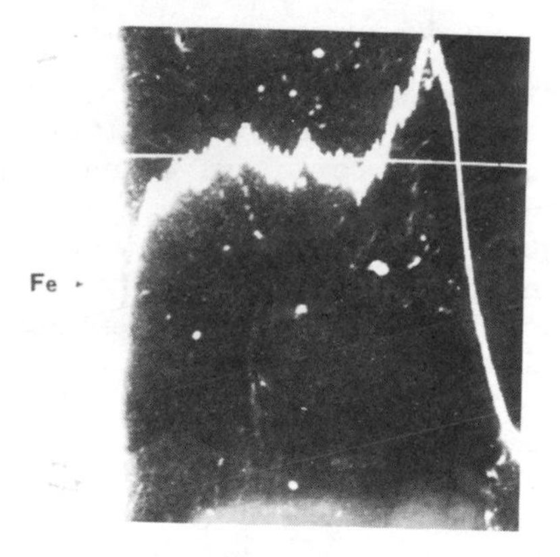

Fig. 6 Chlorine content trend along the section of the film obtained by electrodeposition of resin 134 on 135

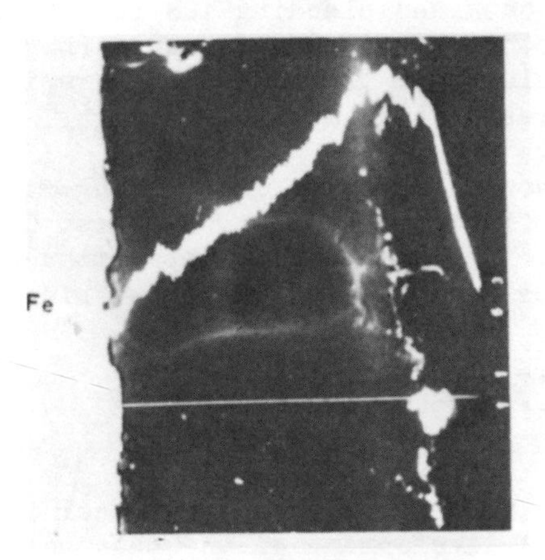

Fig. 7 Chlorine content trend along the section of the film obtained by contemporaneous electrodeposition of resin 128 and 134

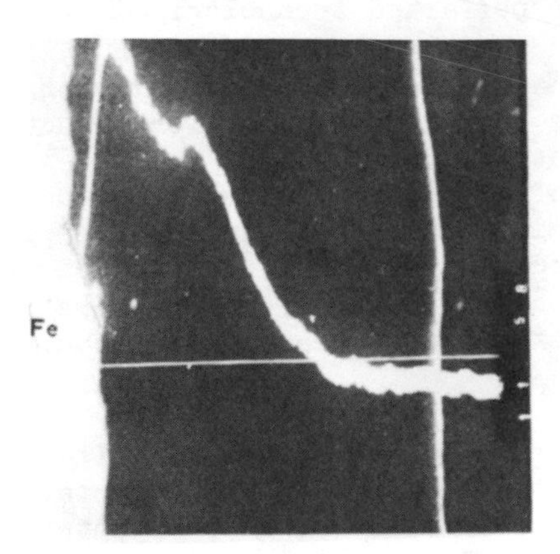

Fig. 8 Chlorine content trend along the section of the film obtained by contemporaneous electrodeposition of resin 127 and 135

The first one was obtained blending 128 resin (low molecular weight, chlorine containing) and 134 resin (higher molecular weight, without chlorine); the second bath resin 127 again toghether with 135 resin (high molecular weight, chlorine containing).
Figures 7 and 8 show chlorine position along films obtained from these baths. These figures show that, at first low molecular weights deposition takes place. The deposition of further materials, splits these fractions on the outer side.

4. TECHNOLOGICAL CHARACTERISTICS OF THE PROCESS

Considered all the foreseeble advantages, car industries got very concerned to the development of the cataphoretic process and started following with more and more interest the improving of the various products supplied by the paint manufacturers.
FIAT laboratories considered the technological characteristics of four cataphoretic products from different manufacturers and worked on each of the following every improvement and any up-to-date solution. The performed technological tests had a double aim:

a - To consider the comparative behaviour among the cataphoretic paints and the anaphoretic one which is usually in production with the purpose to test out the physico-technological characteristics of it, either for what is concerning the stability of the paint bath, or to define the plant changes able to allow a good reliability of the process with such material

b - To choose, among the considered cataphoretic paints, the ones having the best physico-technological characteristics.

4.1 Laboratory tests

- Resistance to corrosion. We performed salt-spray test with different exposure times and scab corrosion tests either on degreased steel-sheet samples or on phosphated ones (the phosphating cycle used in FIAT is zinc-manganese at room temperature) comparing the experimented cataphoretic products with a conventional anaphoretic one.
  In the figures 9 and 10 there is a synthesis of such results. From the performed laboratory tests it came out a good cataphoretic product gives an improvment of 100% on average in resistance to corrosion compared with a conventional anaphoretic one.

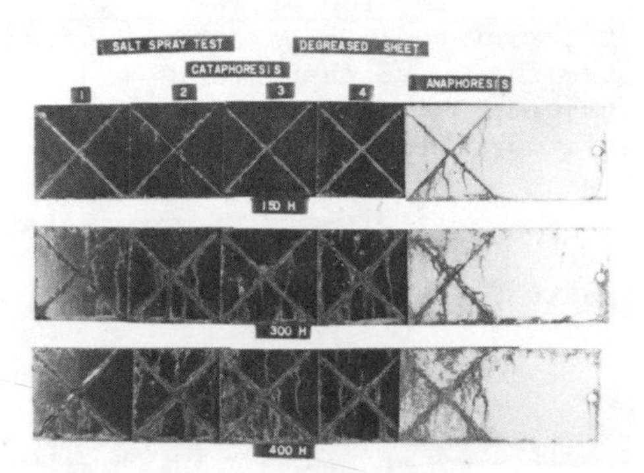

Fig. 9 Comparative salt-spray corrosion test between anaphoresis and cataphoresis applied on degreased steel sheet

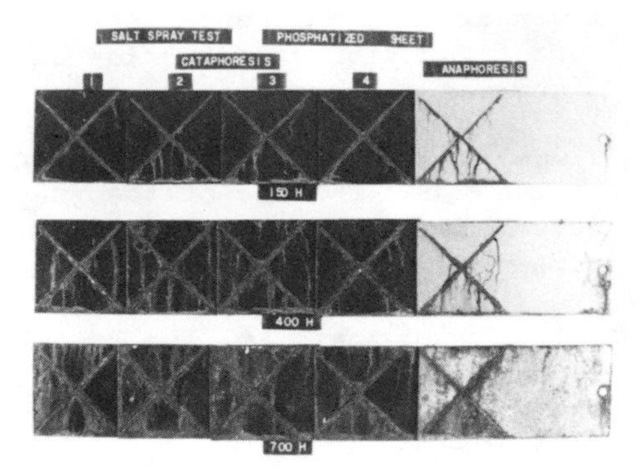

Fig. 10 Comparative salt-spray corrosion test between anaphoresis and cataphoresis applied on phosphated substrate

-Throwing power as a function of resistance to corrosion. We determined the throwing power of cataphoretic products in comparison with the one of an anaphoresis widely used in Europe. The steel-sheet samples, only degreased, on wich we carried out such a test were furthermore exposed for 96 hours to the salt-spray so that we could value comparatively the least film thick ness able to give a sufficiently fair resistance to corrosion. In Fig. 11 there is the respective behaviour of the four exa - mined cataphoretic products compared with the anaphoretic one.

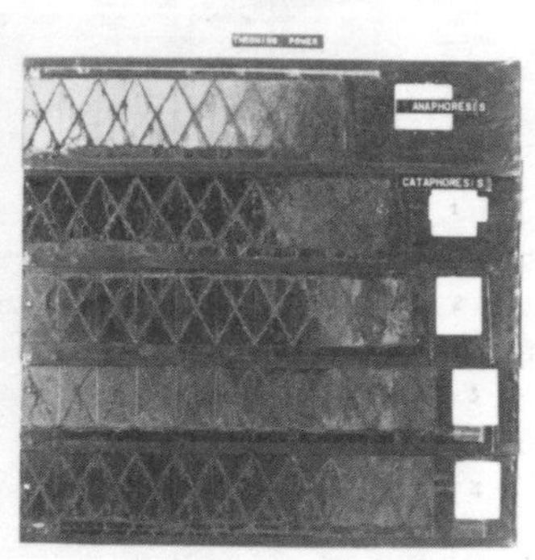

Fig. 11 Comparative salt-spray corrosion test as a function of the throwing power between anaphoresis and cataphoresis

It can be noted that the throwing power of a good cataphoretic paint is quite higher than the one of the anaphoretic product. We can find however that, in order to obtain the same resistance to corrosion, a cataphoretic film needs only half tickness of the one necessary for an anaphoretic film.

-Mechanical resistances of the film. The cataphoretic products generally show a lower elasticity compared with the anaphoretic ones, so that it can be noted, in particular on the complete painting cycle (electrophoretic primer, filler primer, topcoat) a poor adhesion to the phosphate base in drawing and bending tests. We can find a fair improvement curing at least for 30 minutes at a temperature not lower than 180°C. Working on the phosphate base the results are quite improving.

-Influence of the pretreatment quality. Among the various possible ways to improve the adhesion of the cataphoretic products to the phosphate base, one solution could consist of adding a grain refining to the degreasing and phosphating solution.

A more radical intervention consists of doing a heavy phosphating with iron phosphate. As it generally concerns the improvement of the resistance to corrosion we noted that the chromic passivation is very important; in order of importance it is followed by the organic passivation (having ecological characteristics) and the one with trivalent chromium. Anyway a passivation after the phosphating is indispensable.

- Application test. At first some cataphoretic products showed a trend towards the microporosity due to the embedding of microfoam during the dipping of the samples in the painting bath. Furthermore the wet film, after the emersion from the electrophoretic bath, showed some little crumbs; such a defect was eliminable by washing with ultrafiltred solution during the emersion of the piece. The following experiments allowed us to overcome these problems.
- Comments. Referring to the above mentioned advantages and to the promising results verified in laboratory, in FIAT we went on ,starting from November 1977, with the industrial tests on a painting plant for details in order to consider the behaviour of the material during a continue application and to analyze the plant charges necessary for the eventual introduction of this new process for the painting of car bodies

4.2 Test on production plant for steel-sheet parts

The test, started on November 1977, gave very satisfactory application results. As an example we set in Fig. 12 the reproduction of two painted parts, the former with cataphoretic one, exposed for 700 hours in salt-spray. It stands to reason the different of resistance to corrosion shown by the two parts in question. Within the performed test we painted also elements of car body, either by cataphoresis or anaphoresis. Such elements, after having been painted with the complete cycle (filler primer,topcoat) were exposed to environmental resistance tests. In particular we painted with such cycle some

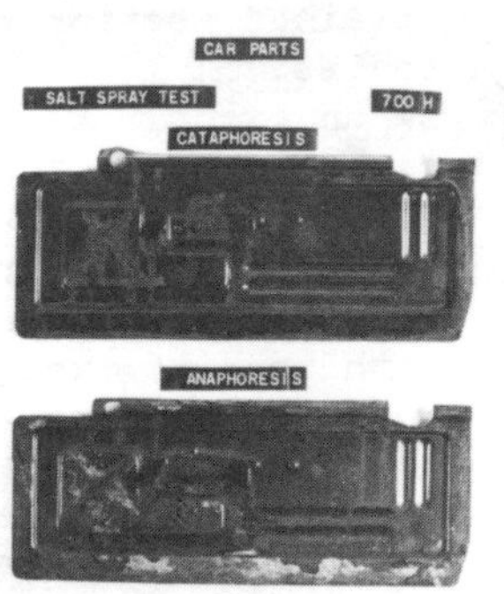

Fig. 12 Comparative salt-spray corrosion test between cataphoresis and anaphoresis applied on car details

Fig. 13 Car doors painted with complete cycle on cataphoretic and anaphoretic primer and exposed for 1000 hours to salt-spray and 3 months to external environment.

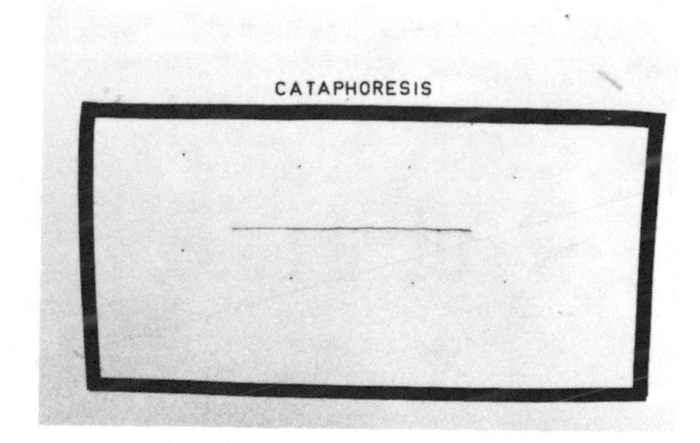

Fig. 14 Detail of fig.13. Door painted by cataphoresis and scratched

car doors, treated as it follows:

- scretch of the film and exposure to salt-spray for 100 hours.
- Following exposure to the external environment for three months.

There results obtained are gathered up in the photos of Fig. 13 - 14 - 15, which show general wiew of two doors with the two different kinds of electrophoretic primer (with filler primer and topcoat) and the details of them near the scratch.

In order to evaluate the throwing power and the resistance to corrosion inside the boxed sections we painted, with cataphoretic and anaphoretic products, some underdoor side frames obtained sectionizing car bodies.

As an example we show in Fig. 16 - 17 - 18 - 19 four meaningful photos of two sides frames painted by cataphoresis compared with ones painted by anaphoresis, exposed for 700 hours in salt-spray and afterwards sectionized.

The two side frames with the cataphoresis were painted without using additional electrodes; the former of the two ones with anaphoresis was painted too without such electrodes,

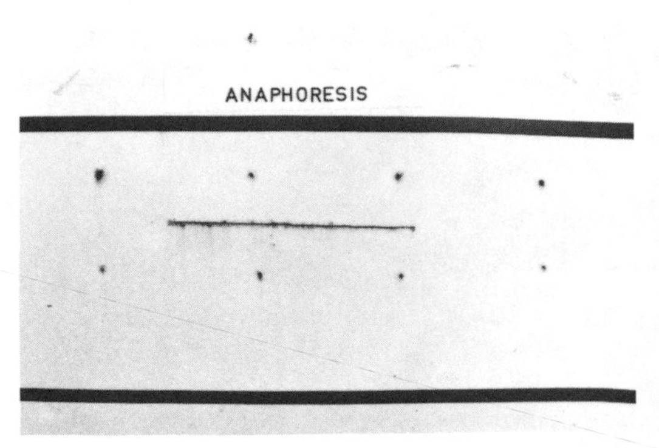

Fig. 15 Detail of fig. 13. Door painted by anaphoresis and scratched

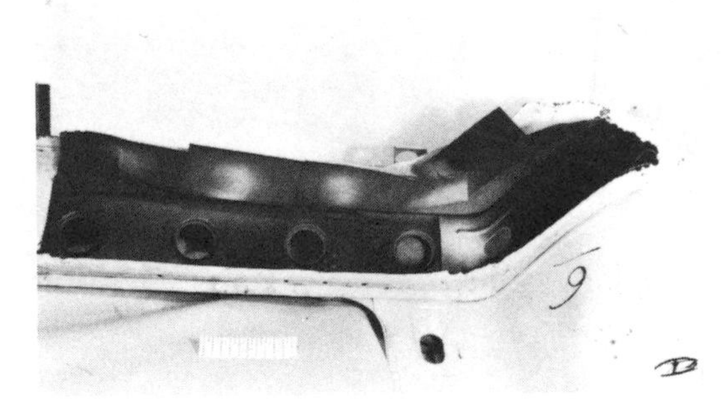

Fig. 16 Detail of underdoor side frame painted by cataphoresis without additional electrode. Thick. 14÷16 μm

while with the latter we used an additional electrode.
The side frames painted by cataphoresis show inside a uniform film thickness (14 μm), while the one painted by anaphoresis has not uniform thickness and show noteworthy tracks of corrosion. The side frame painted by anaphoresis with the additional electrode has a film thickness (28÷30 μm) more or less double than the one painted by cataphoresis, and notwithstanding, it shows a not completely satisfactory resistance to corrosion in the jointings.
Seen these promising re - sults and considered that we did not find anomalies of plant and of painting bath conduction, FIAT decided to carry on this new cataphoretic painting process on a industrial plant for car bodies.

4.3 Industrialization of the process

This new industrial painting process for car bodies started on 1st September 1978 with the filling up of a tank containing 180000 lt. and having a productive potentiality of 800 car bodies day. With the cataphoretic method we paint at present the cars

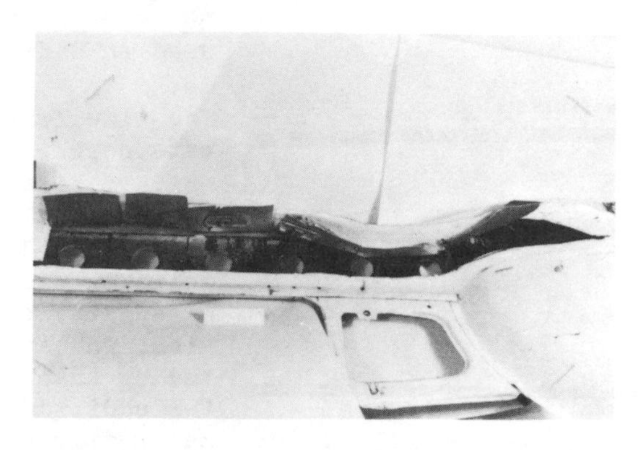

Fig. 17 Detail of underdoor side fra
me painted by anaphoresis
without additional electrode.
Irregular thickness 3÷5 μm

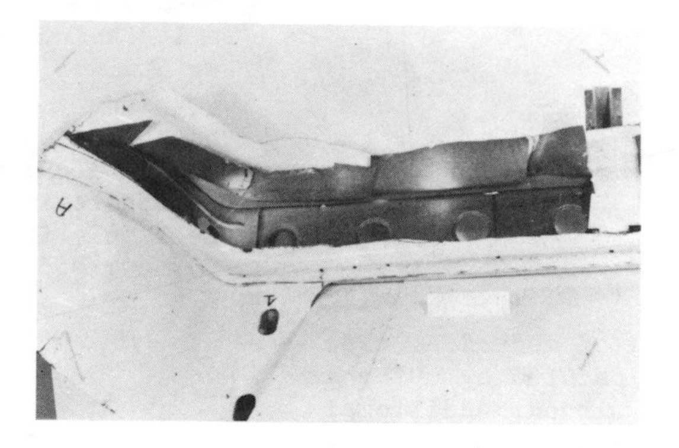

Fig. 18 Detail of underdoor side fra
me painted by anaphoresis
with additional electrode.
Thickness 28 ÷ 30 μm

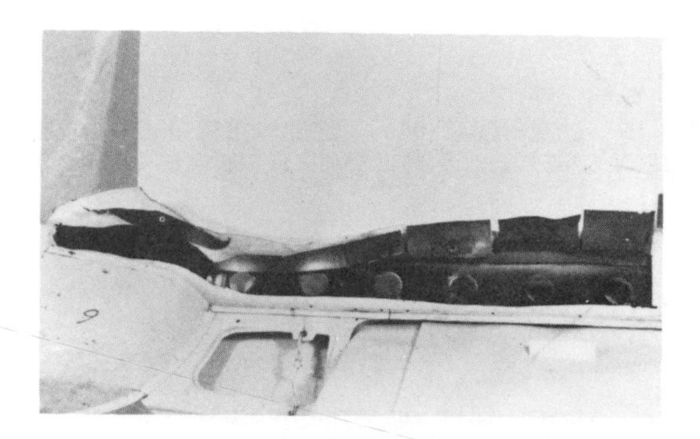

Fig. 19 Detail of underdoor side frame painted by cataphoresis without additional electrode. Thickness 14 ÷ 16 μm

of the new model Ritmo. Now it is too early to weigh in practice the pros and cons of the cathodic painting process, but we can already expect with such a method a higher cost, due to the incidence of the higher cost of the material.
If the already obtained results go on to be confirmed by the test on the industial plant, FIAT will extend this process to all the models.

References

(1) P. L. Bonora, R. Calvillo, Trombetti, G. Bianchini:
'The electrochemical processes of the electrodeposition of resins on metals - 2° Cataphoresis'
FATIPEC Gongress 1978 - Budapest

(2) P. Sennet, J. P. Oliver:
'Colloidal dispersion, electrokinetic effects and the concept of Zeta potential'
Chemistry and Physics of Interfaces pg. 75

(3) F. Beck:
'Fundamental Aspects of electrodeposition of Paint'
Progress on Organic Coating n. 3 (1976) pg. 2

(4) F. Beck:
Proc. 8° Interfinish (1972) Basel pg. 263

THE NYE-CARB® COATING -- WEAR RESISTANCE THROUGH SILICON CARBIDE AND ELECTROLESS NICKEL CODEPOSITION

J. Michael Sale
Electro-Coatings, Inc.

For many years, it was felt that the inclusion of hard particles in a metallic coating would offer certain properties of protection to a substrate material that were previously unattainable. The logic behind such a concept is obvious when one considers the effects of adding sand or gravel to a tar or asphalt coating or, perhaps a more refined analogy, tungsten carbide sintered in a cobalt matrix.

The nature of a coating, of course, is to impart to the surface of a material certain characteristics not inherent to the substrate itself. Altering or modifying the coating to change its properties to further enhance the benefits of its use have at times been problems of a monumental nature and the inclusion of any type of particle in a chemically applied metallic coating certainly fits this category. The reason for this is relatively simple and it really comes to a question of tight chemical controls vis-a-vis stability in the plating bath solution itself.

Although it is not the purpose of this paper to enter into a discussion on plating techniques, it is important to point out that the addition of any material to an electroless metallic plating solution can act as a catalyst and cause the metallic ions to precipitate out of solution onto the catalytic surface as a metal. A certain degree of stability must be maintained or everything the solution contacts will plate (autocatalytic decomposition). However, if the bath becomes too stable, the plating process won't initiate. It should not be difficult to imagine then how critical this stability factor becomes when a large volume of particles is introduced into the plating solution. The problem becomes one of maintaining stability of the bath in a narrow enough range so that the desired part will plate and the particles themselves will not.

It wasn't until the early 1960's that it was demonstrated in a laboratory it was possible to codeposit particles and a metal from an electroless chemical solution and it was more than a decade from then that such a coating could be controlled and offered on a commercial scale. This process is now covered by a number of patents (1).

The rest of this paper will be devoted to a refinement of the electroless composite family of coatings known as NYE-CARB®, its properties, and some of its applications. The NYE-CARB® coating is a codeposition of electroless nickel and particles of silicon carbide. Developed and refined by Electro-Coatings, Inc., it was designed to impart a high wear resistance to the other desirable properties of the electroless nickel coating.

Electroless nickel coatings are formed in the same manner as any other electroless plating whereby nickel ions are reduced to nickel in a hypophosphite and water solution. Without regard for the reaction mechanism and considering only the principal reactants and products, the following equations describe the process:

$$(H_2PO_2)^- + H_2O \xrightarrow{\text{Catalyst}} H(HPO_3)^- + H_2 \quad\ldots\ldots\ldots\ldots \quad (2)$$

$$Ni^{++} + (H_2PO_2)^- + H_2O \xrightarrow{\text{Catalyst}} Ni^\circ + 2H^+ + H(HPO_3)^- \ldots\ldots\ldots\ldots$$

The resultant material is a supersaturated solid solution of nickel and approximately 8% phosphorus. Slight variations in the phosphorus give the coating a laminar appearance but the most unique feature of the material is that prior to heat treating the material is virtually amorphous (its structure is microcrystalline with grains of less than 100 Angstroms in size) and literally pore free. The substrate is encapsulated (3).

The unique feature of the process is its uniformity. Since the process is not dependent on any type of electrical potential, the plating rate at any point in contact with the solution is the same as any other point in the solution. Items that plague other plating processes such as inside diameters, threads, sharp corners, recesses, etc. ... are not factors and the thickness of the final coating is uniform.

Typically, electroless nickel coatings will be heat treated. This step has the effect of altering the grain structure and, consequently the hardness, but the coating still maintains its features of uniformity and lack of porosity. As shown in Figures 1 and 2, electroless nickel can be age hardened to a fairly hard structure and subsequent aging or elevated temperatures will reduce the hardness somewhat (4). The mechanism involved is a reversion from a metastable state to one involving an equilibrium phase mixture of Ni solid solution (containing very little phosphorus) and the intermetallic compound Ni_3P. Warren G. Lee of General American Transportation states, "In our interpretation, the final maximum hardness obtained represents the maximum conversion in the reaction

$$3Ni + P \longrightarrow Ni_3P$$

The dropping off in hardness after the maximum represents a process of slow crystal growth resulting in increased ductility and corrosion resistance of the coating (5)."

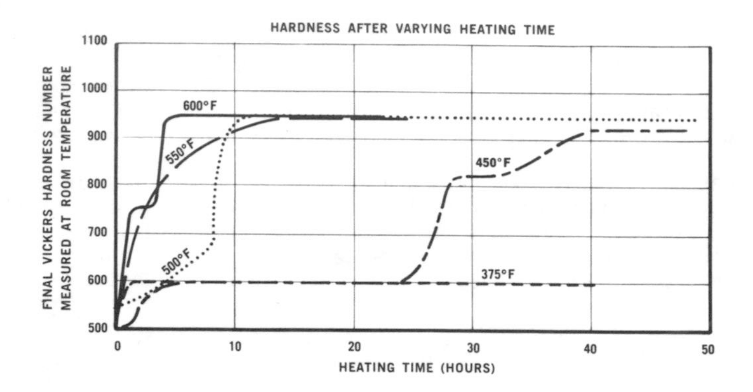

Fig. 1

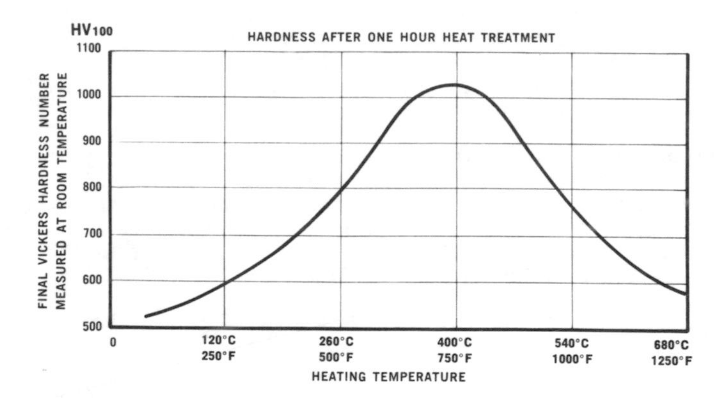

Fig. 2

An electroless nickel deposit renders to its substrate a corrosion resistance at least equal and, in many instances, superior to pure or electro-deposited nickel. Although each corrosive material and environment should be considered individually, certain generalities can be made:

1. Strong solutions of nitric, monochloroacetic, concentrated chromic, sulphuric, and hydrochloric acids, and sodium hypochlorite will attack or complex pure nickel as well as nickel phosphorus alloys (6).

2. The effects of all other corrosive materials are functions of ambient temperatures, motion (i.e., aeration), and the heat treatment of the coating.

Since the NYE-CARB® coating is dependent on electroless nickel as a bonding matrix, it is worth mentioning at this point that electroless nickel has demonstrated bond strengths as high as 60,000 psi (7).

General properties of electroless nickel are found in Table I. The additive to electroless nickel that comprises the NYE-CARB® coating is silicon-carbide in a powder form.

Table I. Properties of Electroless Nickel

Composition	Nickel 90-92% Phosphorus 8-10%
Specific Gravity	7.9
Melting temp. approx.	1635°F (890°C)
Adhesion to Steel	30,000 to 60,000 psi
Hardness (as-plated)	500 to 600 Vickers (49 Rockwell C)
Maximum hardness, heat treated	1000 to 1100 Vickers (67 Rockwell C)
Elongation	2 to 6% permanent strain
Electrical resistance (as-plated)	60-75 microhms/cm/cm^2
Thermal conductivity	0.02 cal/cm sec°C
Coefficient of thermal expansion	13×10^{-6} cm/cm/°C
Reflectivity	50%

Silicon-carbide is an extremely hard material as shown in Table II and is available in many forms. For purposes of the NYE-CARB® coating, particles of 1-3 micron size were selected. The reasoning behind this selection will be discussed later.

Table II. Hardness of Selected Materials

Material	Hardness (Vickers - Kg/mm^2)
Diamond	10,000
Silicon Carbide	4,500
Corundum (AL_2O_3)	2,400
Tungsten Carbide	1,800
Nitrided Steel	1,100
Hard Chrome Plate	1,000 (Rc 70)

The material itself is abundant and easily formed by heating carbon and silica in an electric furnace to 400°F. It is insoluble in water and alcohol, it is also noncombustible and has low toxicity.

In general terms, the silicon-carbide is probably best described as an abundant, innocuous material with a hardness approaching that of diamond. It follows, therefore, with such properties it should probably be capable of demonstrating extremely high resistance to wear at a cost effective level.

The NYE-CARB® coating was devised to combine all of the desirable properties of electroless nickel with those of silicon-carbide. The two materials are compatible and their properties only complement one another and create a wear and corrosion resistant coating which has proven extremely beneficial and protective in many applications. Although other combinations are possible (i.e., electroless nickel and diamonds, electroless cobalt and corundum, etc.) the combination of electroless nickel and silicon-carbide has proven to be the most cost effective and practical. The selection of these materials was not random.

The mechanism for combining electroless nickel and silicon-carbide is relatively simple. The particles are suspended in the plating solution and are more or less bonded as an aggregate as the nickel plating is formed. There is nothing more than a mechanical bond involved.

To obtain the desired properties of this cermet, certain amounts of thought had to go into particle content, particle size, and matrix hardness and/or ductility, the latter being a function primarily of post-coat heat treatment.

Particle concentration in the compound dictates the hardness and, therefore, resistance to certain types of wear. Since silicon-carbide is obviously so much harder than electroless nickel, it follows that the higher the concentration, the higher the hardness. However, each particle must be firmly in place for the coating to be effective and to prevent sloughing. In this regard, a concentration of 25% by volume has proven effective.

Particle size can have various effects on the coating. These range from the coefficient of friction to cosmetics. As particle size increases, the coating darkens to a blue gray and friction coefficients will increase somewhat (this will be discussed later). Surface finish characteristics will be covered in detail; suffice to say here that generally speaking, composites with large particles are difficult to "flow" and subsequently cannot be brought to high lustre finishes. Obviously, at the other extreme, particles too small would be ineffective. Therefore, a NYE-CARB® coating will typically contain particles that are 1-3 micron in size.

To fairly evaluate the properties of NYE-CARB®, the mechanism of the coating should be described. The wear surfaces of silicon-carbide jut out above the nickel matrix. Theoretically, these are the areas exposed to wear and the nickel is only in the recesses or valleys. The photomicrograph in

Figure 3 illustrates this. Therefore, electroless nickel not only provides the encapsulating, corrosion resisting layer as previously described, but it also is the strong, yet relatively ductile, basis material supporting the silicon-carbide particles.

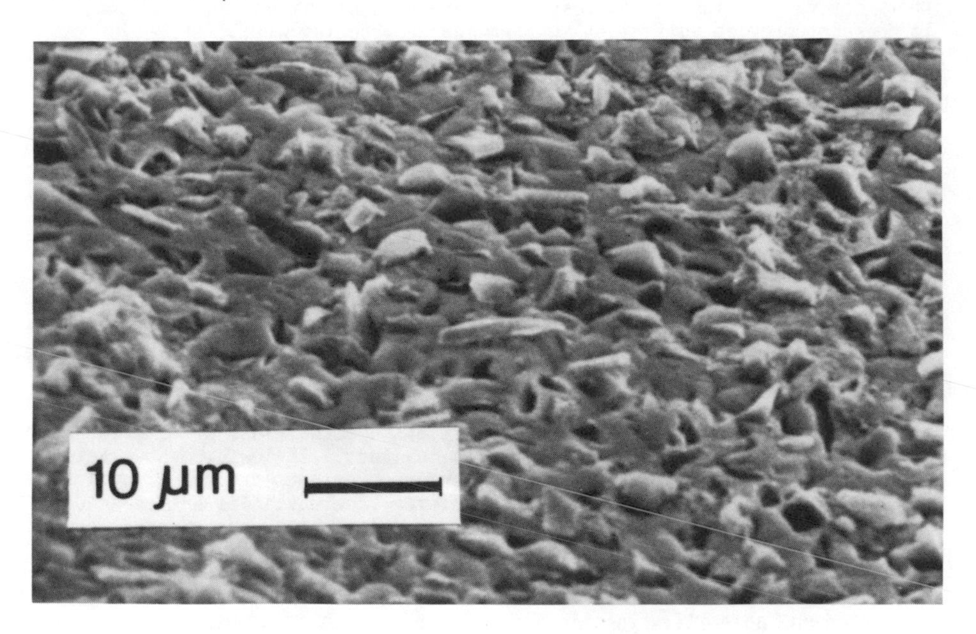

Fig. 3

Since NYE-CARB® has been designed as a wear resistant coating, it would be appropriate to classify wear and its factors to study the NYE-CARB® coating's ability to withstand certain environments.

Howard S. Avery, with his definition of wear as "deterioration due to use", has broken wear into six major causative factors. These are impact, abrasion, friction, heat, corrosion, and vibration (8). Each of these factors will be examined.

Impact. This type of wear creates the condition whereby there is an instantaneous set of compressive and tensile stresses involved at the surface of the material. The compressive forces are focused at the impact point and the rest of the forces involved would be in tension. When the elastic limit is exceeded, deformation in the impact zone will occur and, depending on ductility, subsequent cracking. This is typically a lateral flow phenomenon, but very much dependent on the compressive strength of the material.

The NYE-CARB® coating has a somewhat reduced ductility compared to pure electroless nickel due to the particle concentration. By the same token, its compressive strength is greatly enhanced, however, which tends to compensate to some extent. The substrate material is a real factor here. If it is not fairly high in compressive strength, a situation can occur whereby it does not provide sufficient support and the coating will fail.

In general, any thin coating is poor in applications involving high impact. The lateral forces are simply too high if the compressive forces are not equally spread. NYE-CARB® is a thin coating and should be distinguished from other wear resistant thick coatings such as those applied by welding or metal spraying techniques.

In many applications where the compressive forces are equally spread, the NYE-CARB® coating has worked quite well. One such application was with a drop forge die at Ford Motor Company. This die insert was used to forge connecting rods. The operation involved placing a 2340°F bar on the die and striking it with a 4,000 lb. hammer. Typically, the die would be ruined and scrapped after 15 shifts. Hard chrome plating only lasted less than one shift. A die was plated with the NYE-CARB® coating, however, and it was still intact after 15 shifts.

Abrasion. Wear from this factor results in actual loss of material and it can be categorized based on the forces involved. In low stress abrasion, the abrasive particles rub against the surface but loads are not significant enough to crush them. In high stress abrasion, the particles are crushed. Finally, the worst level is seen as gouging or galling, whereby chunks of metal are sheared out of the substrate.

Abrasive test data on the NYE-CARB® coating using a Taber Abraser are presented on Table III. This is significant in that it compares the NYE-CARB® coating against the matrix material itself and it also points out a considerable difference in particle size and matrix hardness.

A good example of the NYE-CARB® coating's ability to withstand low stress abrasion is in the plastics industry. According to an industry magazine article, the material works well in applications where it is abraded by glass-filled resins (9). Diamond Tool & Die of Dayton, Ohio uses it in a mold for a 30% glass-filled polypropylene wash-armguard for a dishwasher. The silicon-carbide particles prevent the glass from contacting the surface of the mold.

Another application has been on various yarn handling components used in the textile industry. Here, the particles of silicon-carbide suspend the yarn which is traveling at high velocities and prevent it from abrading (at times literally "cutting") the metal surface. This is done without snagging and destroying the yarn.

Table III. Taber Wear Test

Type of Particle	Particle Size	Particle Hardness (Knoop)	Hardness of Matrix Material (Nickel Phosphorus)	Taber Wear Index
None	--	--	49-50 Rc	10.75
None	--	--	65-68 Rc	3.75
Silicon-Carbide	10-15 Micron	2480	49-50 Rc	2.6-3.6
Silicon-Carbide	10-15 Micron	2480	65-68 Rc	1.5-1.9
Silicon-Carbide	1-3 Micron	2480	49-50 Rc	2.9-3.3
Silicon-Carbide	1-3 Micron	2480	65-68 Rc	1.7-2.1
Boron-Carbide	7-10 Micron	2600	49-50 Rc	2.1-2.3
Boron-Carbide	7-10 Micron	2600	65-68 Rc	1.0
Tungsten-Carbide	5 Micron	1800	49-50 Rc	3.0
Tungsten-Carbide	5 Micron	1800	65-68 Rc	2.0

Taber Wear Index = Weight loss in Mg/1000 cycles using CS-10 wheels - 1000 Gm Load - Average of 4000 Cycles. (The lower Taber Wear Index numbers indicate the best abrasion resistance.)

The same mechanism is involved in many high stress abrasion applications. An application that is very severe is found in foundries where sand is blown under pressure against mold or pattern surfaces and then compressed. Again, in this application, the silicon-carbide particles prevent contact and subsequent abrading at the metallic surface.

Gouging such as a grinding environment presents a different problem for the NYE-CARB® coating and, as in the case of impact wear, the substrate hardness will very definitely determine the success or failure of the application. A NYE-CARB® coating on aluminum, for instance, is a poor protection for this type of wear. Forces at the surface will cause a micro-flexure of the coating if it is not adequately supported and consequently, the nickel will be exposed and removed slightly. That will cause a deterioration in its bond of the silicon-carbide. The silicon-carbide particle is released eventually exposing more nickel and so on. Eventually, the coating will be gouged off exposing substrate material. Obviously, severe gouging would present an impossible situation whereby the relatively thin coating would fail rapidly if not immediately.

On a positive note, however, two points must be considered. The first is that the NYE-CARB® coating may provide enough protection for the life of the soft substrate. Secondly, if the substrate itself is sufficiently hard enough to prevent the micro-flexure at the surface (but not hard enough to prevent gouging), the NYE-CARB® may completely alleviate the problem.

Friction. This type of wear can also be described as adhesive wear. Without entering into a complete discussion on the topic, it is sufficient to say that this type of wear occurs when the surfaces of two parts come in intimate enough contact that a welding or an atomic bond actually occurs. When this happens, the parts, as they move, literally tear at one another. Lubrication will prevent contact, reduce this bonding tendency, and subsequently, reduce wear.

A common laboratory test for wear and friction is the LFW-1. Data for this is presented in Table IV. Further friction coefficient data is presented in Table V. In both these examples of data, it is extremely important to point out the need for lubrication.

Although, as in abrasion considerations, the NYE-CARB® coating is very resistant to galling, it can easily gall itself or another material being run against it without proper lubrication. As in most instances of two similar metals run against each other unlubricated, the propensity of nickel to form small atomic bonds is high and, therefore, it will gall. If a NYE-CARB® coating is run against a dissimilar metal unlubricated, the chance for abrasive wear exists due to the particles of silicon-carbide.

With lubrication, as the charts illustrate, the NYE-CARB® coating can significantly reduce friction coefficients and enhance material life. NYE-CARB® enjoys a large number of applications reducing wear from friction. The phosphorus content of the electroless nickel is felt to be contributory, but a major factor is undoubtedly the coating's ability to wet and hold a lubricant. Among the many popular applications are reciprocating pump components, deep draw dies, cam shafts, etc.

Heat. Wear due to heat or thermal cycling is a phenomenon for which the NYE-CARB® coating offers little or no protection. As described earlier, temperatures in excess of 750°F will soften electroless nickel, and above 1000°F the material is not felt to be strong enough to adequately bond the silicon-carbide particles. There are certain liquidus phases at temperatures not too much higher.

Corrosion. Corrosive wear in metals manifests itself in two different forms. The first is by the material actually entering a solution or dissolving and the second, and more insidious, is intergranular. With intergranular corrosion, there is little or no material lost, but the metallic grains or crystals in a material literally become disassociated due to the corroded areas at their boundaries. The subsequent loss of strength, hardness, and resistance to the other forms of wear is substantial, if not catastrophic.

The resistance of a NYE-CARB® coating to this type of wear is typically identical to an electroless nickel coating as described earlier. If other types of wear mechanisms or factors are involved, such as abrasion, it is very advantageous to apply this composite, however.

Table IV. LFW-1 Friction and Wear Test

Test Ring		Test Block		Friction Coefficient				Weight Loss (Mg)	
				Static		Kinetic			
Coating	Surface Hardness	Coating	Surface Hardness	Initial	Final	Initial	Final	Ring	Block
Electroless Nickel	500 Kg/mm^2 Vickers	None	27-33 Rc	--	--	0.11	0.09	102.0	0.2
Electroless Nickel	870 Kg/mm^2 Vickers	None	27-33 Rc	--	--	0.11	0.11	1.0	0.0
Silicon-Carbide Composite	1300 Kg/mm^2 Vickers	None	27-33 Rc	--	--	0.13	0.13	0.1	5.1
Silicon-Carbide Composite	1300 Kg/mm^2 Vickers	Electroless Nickel	870 Kg/mm^2 Vickers	0.16	0.13	0.1	0.1	0.0	6.4
Silicon-Carbide Composite	1300 Kg/mm^2 Vickers	Hard Chrome	1000 Kg/mm^2 Vickers	0.15	0.10	0.07	0.07	0.5	1.1
Silicon-Carbide Composite	1300 Kg/mm^2 Vickers	Silicon-Carbide Composite	1300 Kg/mm^2 Vickers	0.1	0.07	0.09	0.06	0.2	0.2

Test Conditions

Load - 68 Kg.
Speed - 72 RPM
Duration - 5000 Cycles

Lubricant - Mineral Oil
Ring Material - SAE 4620 (ASTM STD D2614-68)
Block Material - SAE 01 (ASTM STD D2614-68)

Table V. Friction Coefficients (Lubricated) - Various Materials

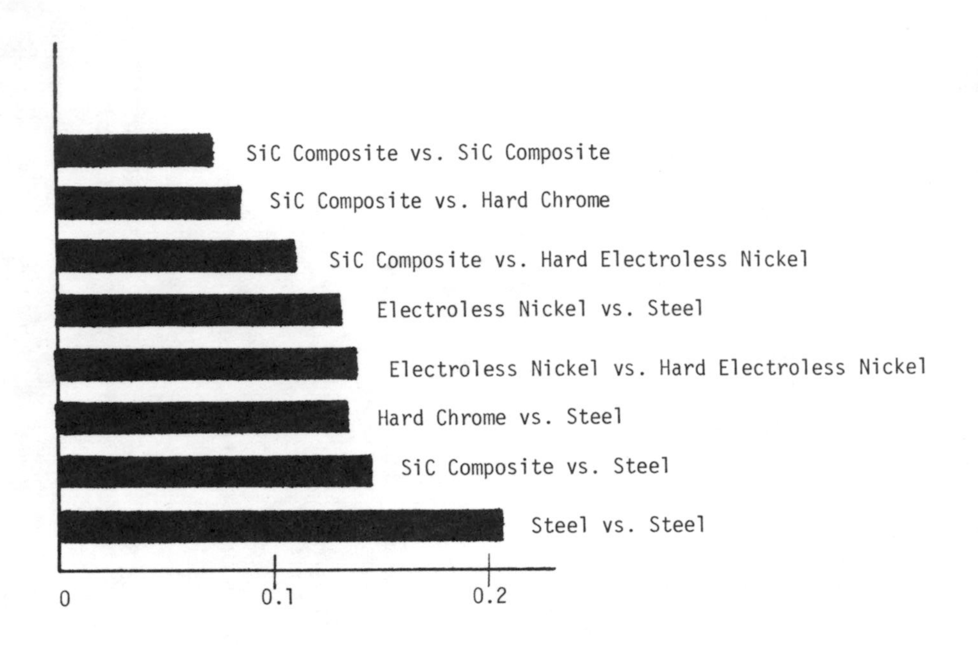

As an example, in the molding of a glass or talc filled polyvinylchloride (PVC) resin, hydrochloric acid is formed. The electroless nickel not only provides the matrix for bonding the silicon-carbide which will be resistant to the abrasion of the filler, it also provides a resistance to the HCl. Since the electroless nickel is also devoid of any refined grain structure, intergranular corrosion is not a factor.

Vibration. This factor can create conditions of surface wear common to both friction and abrasion. It is manifested in a surface phenomenon often described as fretting. The application of a NYE-CARB® coating can provide relief from this type of wear. However, the internal loading and unloading that leads to fatigue can rarely be relieved by a coating. In fact, it should be pointed out at this time that a coating at times can be deleterious to the overall fatigue strength of a material.

Some Other Points

The NYE-CARB® coating is typically applied in thicknesses of .0018" ± .0002" but has been found effective to thicknesses as thin as .0005" in certain applications. This is very much dependent on the substrate material and the operating environment. Thicker applications are possible, but have proven to be of little value.

The surface finish of a NYE-CARB® coating is very dependent on the substrate's finish. Very rarely will the surface finish be as smooth after the NYE-CARB® coating has been applied. Without resorting to any exotic or unusual polishing techniques though, the coating surface can be readily brought to match or exceed the original surface. This is true for finishes up through 1 RMS.

As with electroless nickel, this composite can be applied to most metals. Certain elements in the substrate such as lead or zinc necessitate special pre-treatments. Certain surface treatments will also require special plating methods. These are required on metals that have been nitrided, carburized, anodized, etc. The optimum substrate is steel with a Rockwell C hardness something less than 58.

The NYE-CARB® coating offers wear resistance in application areas where coatings were previously impossible or economically impractical. It is a thin coating and this restricts its use to some extent. However, because of its ability to evenly plate inside diameters, threads, and the most complex of geometries, it is finding new applications, from tiny electrocardiogram pins to 10,000 lb. compression molds. The uniformity so characteristic to a chemically deposited coating, coupled with the many wear and corrosion resistant properties of electroless nickel and silicon carbide, afford the designer latitudes of durability and precision which were previously unattainable.

* * * * * * * * * * * * * *

NYE-CARB® is a registered trademark for a composite electroless nickel and silicon-carbide coating. All patents and rights are held by Electro-Coatings, Inc., 1605 School Street, Moraga, California 94556.

REFERENCES

(1) United States Patents Nos. 3,617,363; 3,753,667; 3,562,000; 3,723,078 (Assigned to Electro-Coatings, Inc.).

(2) Gregoire Gutzeit, "Chemical Reactions", Symposium on Electroless Nickel Plating (Catalytic Deposition of Nickel-Phosphorus Alloys by Chemical Reduction in Aqueous Solution) (Philadelphia: ASTM Special Technical Publication No. 265, 1959), p. 3.

(3) W. H. Metzger, Jr., "Characteristics of Deposits", Symposium on Electroless Nickel Plating (Catalytic Deposition of Nickel-Phosphorus Alloys by Chemical Reduction in Aqueous Solution) (Philadelphia: ASTM Special Technical Publication No. 265, 1959), p. 13.

(4) "Kanigen® Electroless Alloy Coating", General American Transportation Corporation Technical Bulletin No. 874 (1968), p. 5.

(5) Warren G. Lee, "Age of Hardening of Chemical Nickel Coatings", Plating (March, 1960), Copyrighted 1960 American Electroplaters' Society.

(6) G. Gutzeit and E. T. Mapp, "Kanigen® Chemical Nickel Plating", Corrosion Technology, Vol. 3, No. 10 (October, 1956), p. 331.

(7) Metzger, op. cit., p. 15.

(8) H. S. Avery, edited by Albert G. H. Dietz, "Surface by Welding for Wear Resistance", Composite Engineering Laminates (Cambridge, Mass.: The MIT Press, 1969), p. 3.

(9) "New Mold Coating Resists Wear Better", Plastics Technology (June, 1977).